79th Annual Eastern Snow Conference 2023

Easton, Pennsylvania, USA
6-8 June 2023

ISBN: 978-1-7138-8910-6

Printed from e-media with permission by:

Curran Associates, Inc.
57 Morehouse Lane
Red Hook, NY 12571

Some format issues inherent in the e-media version may also appear in this print version.

Printed with permission by Curran Associates, Inc. (2024)

Eastern Snow Conference
C/O Dr. Krystopher Chutko
117 Science Pl-Dept. Geography
Saskatoon, Sk, Canada S7N 5C8

https://www.easternsnow.org/

Additional copies of this publication are available from:

Curran Associates, Inc.
57 Morehouse Lane
Red Hook, NY 12571 USA
Phone: 845-758-0400
Fax: 845-758-2633
Email: curran@proceedings.com
Web: www.proceedings.com

Proceedings of the
79th ANNUAL

EASTERN SNOW CONFERENCE

6-8 June 2023

Nurture Nature Center, Easton, Pennsylvania, USA

Dr. Joshua King
1984-2023

The Proceedings of the 79th Eastern Snow Conference is dedicated to snow scientist and ESC Executive Member Joshua King, who passed away on February 21, 2023 in Toronto, Ontario, Canada, at the age of 38.

CONTENTS

Poster Session #2

Oral Session #4

Oral Session #5

FOREWORD

This proceedings volume contains papers presented at the 79th Eastern Snow Conference (ESC) held 6-8 June 2023 at the Nurture Nature Center in Easton, Pennsylvania. The meeting featured oral and poster presentations covering a wide array of cryospheric science, including but not limited to the remote sensing of lake ice phenology, snow water equivalent estimation in forested and tundra environments, snow satellite mission advancements, and the observation and modeling of snow processes.

The ESC is a joint United States and Canadian forum for discussing recent work on operational, applied and scientific issues related to snow and ice. It also retains an increasing interest as a symposium where novel approaches to cryospheric science of international significance are presented. The ESC has published an annual series of proceedings since 1952. Typical topics include studies of snow and ice as materials, snow removal, meteorological forecasting, river ice control, snow hydrology, snow chemistry, glaciology, remote sensing of snow and ice, and snow ecology. Membership in the ESC is open to all interested individuals and corporations. Additional copies of the current proceedings and all back issues can be obtained from the Secretary. Additional information about the Eastern Snow Conference may be found at: http://www.easternsnow.org/.

We thank members of the ESC executive committee for their continued work toward making each annual meeting a success. This includes, but is not limited to, the work of the Research Committee who evaluate all student presentations and recommend awardees for the three student prizes.

The 2023 meeting of the Eastern Snow Conference and these Proceedings were made possible in part by Corporate membership from Campbell Scientific Canada, Geonor, Gradient Wind Engineering, and Hoskin Scientific Ltd. Printing of the Proceedings was supported by the Centre for Hydrology, University of Saskatchewan.

Gradient Wind Engineering
Ottawa, Ontario, Canada
www.gradientwind.com

GRADIENTWIND
ENGINEERS & SCIENTISTS

Centre for Hydrology
University of Saskatchewan
Saskatoon/Canmore, SK/AB
Canada
https://research-groups.usask.ca/hydrology/

We look forward to seeing you at the 80th Eastern Snow Conference meeting to be held 6-8 June 2024 at the University of Waterloo in Waterloo, Ontario, Canada.

Krystopher Chutko and Craig Smith
ESC Proceedings Co-Editors
University of Saskatchewan and Environment and Climate Change Canada

STATEMENT OF PURPOSE

The Eastern Snow Conference (ESC) is a joint Canadian/US organization founded in the 1940s, originally with members primarily from eastern North America. Our current members are scientists, snow surveyors, engineers, technicians, professors, students, and operational and maintenance professionals from around the world. There is a western counterpart to the ESC, the Western Snow Conference (WSC), which also is a joint Canadian/US organization.

The Eastern Snow Conference is a forum that brings the research and operations communities together to discuss recent work on scientific, applied, and operational issues related to snow and ice. The location of the conference alternates yearly between the United States and Canada, and attendees present their work by either giving a talk or presenting a poster. Most resulting papers are reviewed, edited, and published in our yearly *Proceedings of the Eastern Snow Conference*. In recent years, the ESC meetings have included sessions on snow physics, winter survival of animals, snow and ice loads on structures, river ice, remote sensing of snow and ice, and glacier processes. Volumes of the *Proceedings* can be found in libraries throughout North America and Europe, and the papers are also available through the National Technical Information Service (NTIS) in the United States and CISTI in Canada. All volumes of the Proceedings are now available at *easternsnow.org*.

* * * * * * * * *

Le Colloque sur la neige région Est (ESC) est une organisation américaine-canadienne fondée dans les années '40 et dont les membres provenaient à l'origine principalement de l'est de l'Amérique-du-Nord. Actuellement, les membres, qu'ils soient chercheurs, techniciens, ingénieurs, professeurs, étudiants, et spécialistes des services d'exploitation et d'entretien, viennent non seulement d'Amérique-du-Nord, mais aussi du Royaume-Uni, du Japon, et d'Allemagne. Le Colloque sur la neige-région Ouest (WSC), aussi une organisation américaine-canadienne, est l'homologue de l'ESC pour l'ouest nord-américain.

L'ESC est un forum qui rassemble chercheurs et responsables des services d'exploitation pour discuter des travaux récents sur les problèmes scientifiques, opérationnels, ou autres dus à la neige et à la glace. Le site de cette réunion annuelle alterne entre les États Unis et le Canada. Les participants y présentent les résultats de leurs travaux par des communications orales ou au moyen d'affiches. Ces communications, une fois revues et éditées, sont publiées dans les *Annales de l'ESC*. Dans les années récentes, les réunions de l'ESC ont inclus des sessions sur la physique de la neige, la survie hivernale de la faune, les forces exercées par la neige et la glace sur les structures et les bâtiments, la glace de rivière, la télédétection de la neige et de la glace, et les processus glaciaires. Les *Annales de l'ESC* sont accessibles dans la plupart des bibliothèques scientifiques d'Amérique-du-Nord et d'Europe. Des copies d'articles peuvent être obtenues du National Technical Information Service (NTIS) aux États Unis et son équivalent au Canada, le CISTI. Toutes les volumes des *Annales de l'ESC* sont maintenant disponibles sur *easternsnow.org*.

EXECUTIVES FOR THE 2022-23 79th EASTERN SNOW CONFERENCE

PRESIDENT
Craig Smith
Saskatoon
Saskatchewan, Canada

PAST PRESIDENT
Krystopher Chutko
Saskatoon
Saskatchewan, Canada

VICE-PRESIDENT AND PROGRAM CHAIR
Joan Ramage
Bethlehem
Pennsylvania, USA

SECRETARY-TREASURER (Canada)
Krystopher Chutko
Saskatoon
Saskatchewan, Canada

SECRETARY-TREASURER (USA)
Ken Rancourt
Conway
New Hampshire, USA

Eli Deeb
Hanover
New Hampshire, USA

ESC PROCEEDINGS EDITORS
Krystopher Chutko
Saskatoon
Saskatchewan, Canada

Craig Smith
Saskatoon
Saskatchewan, Canada

***PHYSICAL GEOGRAPHY* EDITORS**
Mauri Pelto
Dudley
Massachusetts, USA

STEERING COMMITTEE
James Brylawski
Augusta
New Jersey, USA

Grant Gunn
Waterloo
Ontario, Canada

Sean Helfrich
Suitland
Maryland, USA

David Robinson
New Brunswick
New Jersey, USA

Laura Thomson
Kingston
Ontario, Canada

Tara Troy
Victoria
British Columbia, Canada

Carrie Vuyovich
Hanover
New Hampshire, USA

RESEARCH COMMITTEE

Sam Tuttle (Chair)
Syracuse
New York, USA

Eunsang Cho
Greenbelt
Maryland, USA

Benoit Montpetit
Ottawa
Ontario, Canada

Melissa Wrzesien
Greenbelt
Maryland, USA

STUDENT REPRESENTATIVES

Mahboubeh Boueshagh
Bethlehem
Pennsylvania, USA

Fraser King
Waterloo
Ontario, Canada

WEBMASTER

Krystopher Chutko
Saskatoon
Saskatchewan, Canada

THE PRESIDENT'S PAGE

The 79th annual Eastern Snow Conference, and our first in-person ESC meeting since 2019, was held in Easton Pennsylvania 6-8 June 2023. It was great to see everyone in person once again, but intertwined with that celebration were somber moments as we mourned the loss of our friend, colleague, and ESC Executive Josh King who tragically passed away on February 21st of this year. It was heartening to see Josh being remembered and honored throughout the meeting, and although it feels miniscule considering what Josh accomplished in his short time as a research scientist, the ESC would like to dedicate these proceedings to his memory. He is missed tremendously by this community.

The 79th ESC was a resounding success, with 54 total registrants, 17 of which were students. On behalf of all the participants, I would like to thank Joan Ramage, our Vice President and Program Chair, and her team of local volunteers for putting together an interesting and diverse scientific program and arranging some fun-filled social events. Congratulations to our student award winners: Jake Ferguson for winning the Wiesnet Medal for best student paper and Vicki Jagdeo for winning the David Miller Prize for best student poster. Well deserved, and best of luck to all of our student presenters with their continued academics and early careers. I hope to see many of you back again next year, and for many ESC meetings to come.

It was my honour to present ESC Life Membership Awards to two very deserving candidates, Miles Ecclestone and Dorothy Hall, and it was very enjoyable to hear them both give keynote addresses at the banquet, sharing stories of their careers that are intertwined with their long and outstanding (and very much appreciated) involvement with the ESC. Life Members Barry Goodison and James Foster were also in attendance.

Always a highlight of the week was the presentation of the coveted Sno-foo award, presented by Benoit Montpetit to our colleague Peter Toose for proudly producing, choreographing, and distributing videography of the ill-fated demise of his research drone. Well done Peter!

With my term as ESC president ended, I would like to extend my gratitude to Past-President Krys Chutko who tirelessly serves as registrar, webmaster, and guidance counselor, alongside his regular duties as Secretary-Treasurer (Canada) and ESC Proceedings Editor. His "corporate memory" keeps the organization running smoothly and the traditions alive.

As much fun as it was to get together in-person in Easton, I think that we should do it all again next June in Waterloo. I have every confidence that our new Vice President, Grant Gunn, with help from his local team and with full support from the ESC Executive, will organize another fantastic meeting.

One last point in closing, I would like to reaffirm the commitment of the ESC to equity, diversity, and inclusion. We aren't always perfect, but it is up to all of us to strive to be better examples for the next generation of snow researchers.

See you next June in Waterloo!

Craig D. Smith
79th President, Eastern Snow Conference

LIFE MEMBERS OF THE EASTERN SNOW CONFERENCE

The Eastern Snow Conference gratefully recognizes individuals who have made lifelong contributions to the study of snow and for their support of this organization.

MILES ECCLESTONE

JAMES FOSTER

BARRY GOODISON

DOROTHY HALL

GERRY JONES

JOHN METCALFE

HILDA SNELLING

DONALD WIESNET

In memoriam

PETER ADAMS

DON DUNLAP

ART ESCHNER

AUSTIN HOGAN

ROBERT B. SYKES

ESC Life Members Barry Goodison, James Foster, Dorothy Hall, and Miles Ecclestone at the 79th Annual Eastern Snow Conference, 2023.

Wiesnet Medal

The Eastern Snow Conference encourages student research through its Wiesnet Medal. This medal is presented annually to the best student paper presented at the conference.

Year	Winner	Affiliation
2023	Jake Ferguson	University of Waterloo
2022	Megan Verfaillie	University of New Hampshire
2021	Daniela Krampe	Alfred-Wegener-Institut
2019	Zhibang Lv	University of Saskatchewan
2018	Jiyue Zhu	University of Michigan
2017	Caroline Dolant	Université de Sherbrooke
2016	Syed Mousavi	University of Michigan
2015	Nicolas Leroux	University of Saskatchewan
2014	Justin Hartnett	Syracuse University
2013	Andreas Dietz	Earth Observation Center/DFD, Germany
2012	Elizabeth Burakowski	University of New Hampshire
2011	Kathryn Semmens	Lehigh University
2010	Simon von de Wall	University of Victoria
2009	Si Chen	Dartmouth College
2008	Chris Fuhrmann	University of North Carolina at Chapel Hill
2007	*Not awarded*	
2006	Y.C. Chung	University of Michigan
2005	M. Javan-Mashmool	Université du Québec à Chicoutimi
2004	J. Farzaneh-Dehkordi	Université du Québec à Chicoutimi
2003	Alexandre Langlois	Université de Sherbrooke
2002	Patrick Ménard	Université de Laval
2001	C. Tavakoli	Université du Québec à Chicoutimi
2000	*Not awarded*	
1999	S. Brettschnieder	Université du Québec à Chicoutimi
1998	Andrew Grundstein	University of Delaware
1997	Newell Hedstrom	University of Saskatchewan
1996	Suzanne Hartley	University of Denver
1995	Paul Wolfe	Wilfred Laurier University
1994	G.E. Mann	University of Michigan
1993	G. Devarennes	Université du Québec à Québec
1992	D.W. Cline	University of Colorado
1991	D. Samelson	Cornell University
1990	A.K. Abdel-Zaher	University of New Brunswick
1989	A. Giguere	McGill University
1988	Mauri Pelto	University of Maine
1987	Cameron Wake	Wilfred Laurier University
1986	Craig Allan	Trent University
1985	Robert Speck	Rensselaer Polytechnic Institute
1984	N.K. Kalliomaki	Laurentian University
1983	David Beresford	Trent University
1982	*Not awarded*	
1981	Jeffrey Patch	University of New Brunswick
1980	Bryan Wolfe	Trent University
1979	Margaret Leech	McGill University
1978	Michael English	Trent University
1977	Don McLaughlin & George Duggan	Rensselaer Polytechnic Institute
1976	Dwayne McMurter	Trent University
1975	Nigel Allan	Syracuse University
1974	*Not awarded*	
1973	Stan Mathewson	Trent University

David Miller Award

The David Miller Award is presented to the best student poster at the annual Conference.

Year	Winner	Affiliation
2023	Vicki Jagdeo	Lehigh University
2022	Mariah Matias	Lehigh University
2021	Madison Woodley	Syracuse University
2019	Gaohong Yin	University of Maryland
2018	Justin Pflug	University of Washington
2017	Elizabeth Ryan	University of Maryland
2016	Oliver Wigmore	The Ohio State University
2015	Charles Papasodoro	Université de Sherbrooke
2014	Laura Thomson	University of Ottawa
2013	Nastaran Saberi	University of Waterloo
2012	Johnathan Sugg	Appalachian State University
2011	*Not awarded*	
2010	Nicolas Svacina	University of Waterloo
2009	Iliyana Dobreva	Texas A&M University
2008	Christopher Karmosky	Pennsylvania State University

Campbell Scientific Canada Prize

The Campbell Scientific Canada prize is awarded to student research showing the most innovative use of technology in the gathering of data.

Year	Winner	Affiliation
2023	*Not awarded*	
2022	Alicia Pouw	Wilfred Laurier University
2021	Fraser King	University of Waterloo
2019	Alex Mavrovic	Université de Sherbrooke
2018	*Not awarded*	
2017	Abigail Dalton	University of Ottawa
2016	*Not awarded*	
2015	*Not awarded*	
2014	Aaron Campbell	University of Waterloo
2013	*Not awarded*	
2012	Andrew Kasurak	University of Waterloo
2011	Mathieu Beaulieu	University of British Columbia
2010	*Not awarded*	
2009	*Not awarded*	
2008	Nicholas Kinar	University of Saskatchewan
2007	Gro Lilbaek	University of Saskatchewan
2006	Gernot Koboltschnig	University of Natural Resources and Applied Life Sciences, Vienna
2005	Warren Helgason	University of Saskatchewan

Oral Session #1

Session Chairs: Laura Thomson & Mahboubeh Boueshagh

Eastern Snow Conference 2020-2022: The Virtual Years

KRYSTOPHER J. CHUTKO[1]

ABSTRACT

Like all organizations, the Eastern Snow Conference (ESC) was directly impacted by the COVID-19 pandemic. And like all other organizations, the ESC adjusted and adapted to the range of real and threatened restrictions imposed by federal, regional, and local governments. Thus, the uninterrupted streak of annual ESC meetings over seven decade (1950's – 2010's) was broken, resulting in no meeting in 2020 and virtual meetings in 2021 and 2022. This new format came with a new set of benefits and challenges, many of which played out behind the scenes and were not publicly announced. This presentation will discuss what was lost and gained during these three unusual years for the ESC. Attention will be paid towards the lessons learned in hosting virtual meetings, participating in a large joint conference with other organizations, and the financial impacts on the non-profit organization of the ESC. The results suggest that the cryosphere sciences community is resilient, persistent, and supportive throughout such challenges, emphasizing the importance of organizations like the Eastern Snow Conference as a leading forum for the community.

[1] Department of Geography and Planning, University of Saskatchewan, Saskatoon, SK, Canada
Corresponding author: krys.chutko@usask.ca

NASA SnowEx 2023 Tundra and Boreal Forest Field Campaign in Alaska, U.S.

CARRIE VUYOVICH[1], SVETLANA STUEFER[2], HANS PETER MARSHALL[3], MICHAEL DURAND[4], KELLY ELDER[5], DRAGOS VAS[6], BATUHAN OSMANOGLU[1], CHRISTOPHER LARSEN[2], DANIEL HODKINSON[1,7], KELLY GLEASON[8], ANNE NOLIN[9], MEGAN MASON[1,7], ARTHUR GELVIN[6], AND ELIAS DEEB[10]

ABSTRACT

NASA's SnowEx was initiated by the Terrestrial Hydrology Program in 2017 to study snow remote sensing challenges in different environments in preparation for a future snow mission opportunity. The specific focus of SnowEx is on testing and maturing technology for satellite remote sensing of global snow water equivalent (SWE) and albedo. Over the 2022-2023 winter, SnowEx an airborne and ground-based snow campaign was conducted in Alaska to address SWE, snow depth and albedo measurement questions unique to taiga and tundra snowpacks. Four sites were selected in Interior Alaska, a boreal forest environment with discontinuous permafrost and seasonal taiga snowpack. Two sites were located on the North Slope of Alaska, a region dominated by low-stature land cover, tundra snowpack, and continuous permafrost. A suite of airborne and ground-based validation activities in fall 2022 and spring 2023 were conducted to quantify and compare the capabilities of radar, altimetry and multispectral sensors to measure SWE and snow depth during the winter season and albedo during the spring melt period. The SnowEx field campaign will provide snow datasets in support of testing and advancement of remote sensing, modeling, and measurements techniques needed for the development of global SWE products. This presentation will provide an overview of this winter's field and airborne activities in Alaska and a summary of the data collected.

---

[1] NASA Goddard Space Flight Center, Greenbelt, MD, USA
[2] University of Alaska – Fairbanks, Fairbanks, AK, USA
[3] Boise State University, Boise, ID, USA
[4] Ohio State University, Columbus, OH, USA
[5] U.S. Forest Service, Fraser, CO, USA
[6] Cold Regions Research and Engineering Laboratory, Fairbanks, AK, USA
[7] Science Systems and Applications, Inc, Lanham, MD, USA
[8] Portland State University, Portland, OR, USA
[9] University of Nevada, Reno, Reno, NV, USA
[10] Cold Regions Research and Engineering Laboratory, Hanover, NH, USA
Corresponding author: carrie.m.vuyovich@nasa.gov

Fig. 1. Snow measurements being made at the Upper Kuparuk-Toolik tundra site during the March SnowEx 2023 campaign, Photo credit: Svetlana Stuefer.

Application of a Deep Learning Nested U-Net for Reflectivity Inpainting in Spaceborne Radar Blind-Zones

FRASER KING,[1] CLAIRE PETTERSEN,[1] AND CHRISTOPHER G. FLETCHER[2]

ABSTRACT

CloudSat's Cloud Profiling Radar (CPR) is one of the few observation-based systems for remotely monitoring high-latitude snowfall. However, the CPR is unable to observe hydrometeor activity within the lowest 1.2 km of the atmosphere due to ground-clutter contamination. This radar "blind-zone" limits CloudSat's ability to detect and measure the intensity of shallow snowfall processes occurring near Earth's surface, leading to increased uncertainty in CloudSat-derived snowfall estimates. Here, we develop a deep learning U-Net++-style convolutional neural network (i.e. blindpaint) for predicting reflectivity within a radar blind-zone. By relating latent information encoded in blind-zone-aloft clouds with additional context from collocated atmospheric climate variables, blindpaint learns to predict the presence and intensity of near surface reflectivities. Blindpaint demonstrates a higher proof of detection and lower false alarm rate to traditional inpainting methods, indicating an improved ability to predict the presence of near surface hydrometeors. The pixel-level accuracy of the inpainted blind-zone reflectivity values from blindpaint also demonstrate mean absolute error performance improvements that are an order of magnitude lower than traditional inpainting techniques. Additionally, by training blindpaint on a combination of CPR-calibrated KaZR surface radar datasets at multiple locations (along with collocated ERA-5 atmospheric data), and simulating the blind-zone, we are able to produce a generalized model which can then be applied to spaceborne observations from CloudSat. Machine learning inpainting techniques like those explored in this work demonstrate a compelling utility in enhancing current and future spaceborne remote sensing missions by revealing important connections between the blind-zone and the surrounding atmospheric state.

---

[1] Climate and Space Sciences and Engineering, University of Michigan, Ann Arbor, MI, USA
[2] Department of Geography & Environmental Management, University of Waterloo, Waterloo, ON, Canada
Corresponding author: kingfr@umich.edu

Modelling and Analysis of Cross-Polarized Radar Backscatter at C, X, and Ku Bands for SWE Retrieval Algorithm

FIROZ BORAH[1], ZHENMING HUANG[1], LEUNG TSANG[1], AND EDWARD KIM[2]

ABSTRACT

There have been many studies of radar backscatter response to snow which include satellite (Sentinel-1), airborne (NASA SnowEx) and tower-based radar measurement (NoSREx) at C, X, and Ku bands and even planned future satellite missions. These studies have been vital in developing and advancing the X and Ku dual frequency retrieval algorithm for SWE (snow water equivalent). Although the primary focus of the algorithm is the co-polarized radar backscatter, the cross-polarized signal can give important details about the snowpack as well. As such, studying and modelling cross-pol radar backscatter from snow correctly and accurately is important.

The cross-pol is predicted by modelling snow as a bi-continuous media of air and ice, controlled by grain size parameter k_c and aggregation parameter b and then solved using the DMRT (dense media radiative transfer). Parameter b is kept small ($b = 0.4$) to have clusters of ice particles which increases the cross-pol. The DMRT equations are solved iteratively, and it can be shown that cross-pol majorly contributes to the higher order terms. Including the backscattering enhancement results in 2.5-3 dB enhancement for co-pol and cross-pol backscatter, which increases the cross-pol to co-pol ratio as predicted by Sentinel-1. The cross-pol radar data is also used to reduce the need for prior information when doing the SWE retrieval using the X and Ku dual frequency algorithm. The cross-pol radar measurements are used to do the classification of the co-pol radar data and this classification is used to drive the algorithm to find the correct solution for SWE. This method is validated using both airborne and tower radar data.

[1] Radiation Laboratory, Department of Electrical and Computer Engineering, University of Michigan, Ann Arbor, MI, USA
[2] NASA Goddard Space Flight Center, Greenbelt, MD, USA
Corresponding author: firozb@umich.edu

Spatiotemporal Prediction of Snow Accumulation with Recurrent Graph Convolutional Networks

BENJAMIN ZALATAN[1] AND MARYAM RAHNEMOONFAR[1]

ABSTRACT

The accurate prediction and estimation of annual snow accumulation has grown in importance as we deal with the effects of climate change and the increase of global atmospheric temperatures. Airborne radar sensors, such as the Snow Radar, are able to measure accumulation rate patterns at a large-scale and monitor the effects of ongoing climate change on Greenland's precipitation and run-off. Tracking and forecasting these internal ice sheet layers is important for calculating snow mass balance, extrapolating ice age from direct measurements of the subsurface, and inferring otherwise difficult to observe ice dynamic processes. Precise understanding of the spatiotemporal variability of snow accumulation in the Greenland ice sheet is important to reducing the uncertainties in current climate model predictions and future sea level rise. Given the amount of snow accumulation in previous years using the radar data, in this paper, we propose a machine learning model based on recurrent graph convolutional networks to predict the snow accumulation in recent consecutive years at a certain location. In this work, we proposed a temporal, geometric, multi-target machine learning model based on GCN-LSTM that predicts the annual snow accumulation of Greenland from 2007 to 2011 given the annual snow accumulation from 1997 to 2006. Our proposed model was shown to perform better and with more consistency than equivalent non-geometric and non-temporal models. While our model succeeds at predicting shallow layer thicknesses with reasonable accuracy, there are opportunities for improvement and generalization.

[1] Department of Computer Science and Engineering, Lehigh University, Bethlehem PA, USA
Corresponding author: maryam@lehigh.edu

Influence of Snow Capture by Forest Canopy for a Seasonal Snowpack in the Adirondack Mountains, NY

MADISON WOODLEY[1], HEATHER GUNN[1], AND SAMUEL TUTTLE[1]

ABSTRACT

During the winters in Upstate New York, seasonal snow is often captured by tree canopies that delay snow from depositing on the surface and affect the rate at which the snow melts in the snowpack. Tree canopies often shade the snowpack slowing sublimation and melting rates if most of the snowpack is covered by trees. Differential melting of the snowpack occurs if the snowpack receives uneven amounts of sunlight from forested areas that have a mix of open and closed canopies. Quantifying the effect of canopy interception of snow is difficult because many snow study sites are focused on open areas and/or fields where instruments, satellites, and manual measurements are more accessible. We focus on the Arbutus Lake watershed, located within the Huntington Wildlife Forest (HWF) in the central Adirondack Mountains of New York, USA. The Arbutus Lake watershed is heavily forested with a mix of conifer and deciduous trees. For our study, we use traditional snow depth/SWE transect methods collected on a biweekly timeframe to quantify how snow depth (SD) and snow water equivalent (SWE) vary between three canopy types: heavily coniferous, heavily deciduous, and an entire open canopy field. We examine the spatiotemporal dynamics of the snowpack under different energy budgets and relate it to midwinter melt events. Preliminary results show a distinct change in SD and SWE values in our snowpack depending on whether it is beneath a closed or open canopy type. Additional results show that SD and SWE values differ depending on whether there has been a recent snowstorm or after a long period of melt.

[1] Department of Earth and Environmental Sciences, Syracuse University, Syracuse, NY, USA
Corresponding author: mwoodley@syr.edu

Poster Session #1

Session Chair: Joan Ramage

Retrieving Snow Density from Ground-Based Radar and Airborne Lidar Observations and Spatial Prediction for Distributed Snow Water Equivalent in Sub-Alpine Mountain Environments

TATE G. MEEHAN[1], SHAD O'NEEL[1], H.P. MARSHALL[2], ELIAS J. DEEB[1], AHMAD HOJATIMALEKSHAH[2], ZACHARY KESKINEN[2], AND BRENT WILDER[2]

ABSTRACT

Constraining snow depth from repeated airborne lidar acquisitions enables the retrieval of snow density from radar travel-time along transects at 1-m resolution. These inferred observations of bulk snow density illustrate spatial patterns and features that are not captured by sparse manual measurements. We combine ground-penetrating radar (GPR) surveys and airborne lidar acquisitions in subalpine environments, covering a range of terrain and ecological parameters such as elevation, slope, aspect, forest canopy, and burned forest area, which captured variability in snow density due to differences in surface energy balance and wind exposure. These terrain and vegetation parameters can be derived from airborne lidar acquisitions and serve as the basis of features within a supervised learning framework to extrapolate snow density estimates across the study regions. Several machine learning models of snow density are compared, but the choice of a best model is difficult to quantitatively assess, as outputs have similar statistical representations but differing spatial patterns. Estimates of the snow water equivalent (SWE) are calculated, using the density map and lidar snow depths, at a scale of tens of square kilometers. The total SWE estimated by any choice of density model is similar in bulk value. This work implores a deepened focus on the meteorological, terrain, and ecological variables controlling the densification of snow in subalpine mountain watersheds with the intent of studying predictor importance and model applicability. An improved observational comprehension of the influences of snow densification will enable snow scientists to better assess and improve physically modeled snow density.

[1] U.S. Army Cold Regions Research and Engineering Laboratory, Hanover, NH, USA

[2] Boise State University, Boise, ID, USA

Corresponding author: tate.g.meehan@erdc.dren.mil

Figure 1. Snow water equivalent of Mores Creek Summit, Idaho was estimated from helicopter-borne lidar acquisitions and density spatially distributed via artificial neural networks.

Implementation and Field Validation of a Passive Radioisotope SWE Sensor in the Catskill Mountains, NY

DAVID KELLEY[1]

ABSTRACT

Snow water equivalent (SWE) is an important environmental variable in studies of the hydrologic cycle, water availability, and climate change. Conventional methods of SWE measurement are labor intensive and can produce errors during freeze/thaw cycles. Automated non-contact samplers can reduce or eliminate these issues. We tested a Campbell Scientific CS725[2] at one site in the upper Neversink River Basin for possible inclusion in the USGS Next Generation Water Observing System. The instrument determines SWE by passively monitoring the attenuation in the snowpack of radiation emitted by potassium and thallium in the soil. To maximize the number of comparable datasets, the instrument was co-located at the long-term Frost Valley YMCA National Atmospheric Deposition Program monitoring site in the Catskill Mountains at Claryville, NY. Data were collected in 6-hour intervals for three seasons. To verify the accuracy of the CS725, conventional measurements of SWE were collected whenever substantial changes in snowpack occurred. Measurements were made at four locations along a 30-meter snow course running through the experimental site. At each location SWE was measured using a Snowmetrics[2] 12-inch tube sampler and a 250cc wedge sampler. The mean SWE was calculated from the conventional measurements and compared to the CS725. The difference between the methods ranged from -16.9 mm to +11.2 mm. In general, the CS725 measurement was 5-20% higher than the conventional method, with a median of 11%, which falls within the 5-35% range found in similar studies of the CS725.

---

[1] New York Water Science Center, U.S. Geological Survey, Troy, NY, USA

[2] Any use of trade, firm, or product names is for descriptive purposes only and does not imply endorsement by the U.S. Government

Corresponding author: dkelley@usgs.gov

Bulk Snow Density Retrievals from Passive Microwave Remote Sensing and Automatic Weather Stations in a Tundra Environment

JEFFREY WELCH[1] AND RICHARD KELLY[1]

ABSTRACT

Available methods to monitor snow density are limited spatially and temporally. *In situ* methods require substantial labour and are relatively costly. This research proposes a novel technique to retrieve bulk snow density estimates from passive microwave (PM) remote sensing. A traditional style PM snow water equivalent (SWE) model is rearranged so snow density is the variable of interest instead of snow depth. Daily automatic weather station (AWS) observations are used to parameterize the snowpack in the scene with snow depth and air temperature observations. The Environment and Climate Change Canada AWS in Eureka, Nunavut is chosen to train and validate the algorithm. Snow density records are available for the Eureka site through the Canadian Historical Snow Water Equivalent dataset. Two snowpack models are investigated for this algorithm – a one-layer model (typically used in PM SWE retrievals) and a two-layer model (a slab layer with underlying depth hoar layer). In the two-layer model, 3D gradient descent is used to produce density estimates for each layer, which are then aggregated for a bulk snow density estimate. Climatological analysis of the site provides some prior knowledge of the conditions of the scene, to further control the behaviour of the machine learning (ML) algorithm. These controls are included as components of the ML's cost function (with the PM signal component of standard ML) to introduce temporality, inertia, and logic into the algorithm. Ultimately, this approach could be expanded to retrieve snow density estimates from PM observations over the Northern Hemisphere.

[1]Interdisciplinary Centre on Climate Change, and Department of Geography and Environmental Management, University of Waterloo, Waterloo, ON, Canada

Corresponding author: jjwelch@uwaterloo.ca

Post Processing Techniques for better Surface Density Estimates for use in Wildlife Tracking Applications

JACK DECHOW[1,2], MICHAEL DURAND[1,2], LAURA PRUGH[2], JESSICA LUNDQUIST[3], BEN SULLENDER[3], CALLUM CUNNINGHAM[3], CASSIE LUMBRAZO[3], AND KATHERINE BREEN[3]

ABSTRACT

Surface layer snow density has a direct impact on the mobility of wildlife in the winter season. There is a density threshold that determines whether or not a snowpack can support the weight of an animal. Thus, the pathing of wildlife throughout the winter is directly impacted by the spatial distribution of snow density in their habitats. We present two methods to better estimate surface density at scale. First, using a known model to determine bulk density from snow depth, we tune the model parameters with least squares adjustment using depth and surface density data from the ABoVE field campaign. Second, we developed a non-linear state model that steps through the modeling time period and incorporates satellite derived Land Surface Temperature (LST) measurements to better simulate snowpack densification resulting from melt-freeze of the snowpack surface on warm days. For the goal of wide scale wildlife tracking across Washington State, we use the NoahMP snow model within the Land Interface System Framework (LISF) to produce initial estimates of snow characteristics (multilayer density, SWE, depth). The appropriate snow characteristics are used as the inputs to both methods and we present a comparison of results and the implications for wildlife tracking purposes in Washington State.

[1] School of Earth Sciences. The Ohio State University, Columbus, OH, USA
[2] Byrd Polar Research Center, The Ohio State University, Columbus, OH, USA
[3] Civil & Environmental Engineering, University of Washington, Seattle, WA, USA
Corresponding author: dechow.5@osu.edu

Passive Microwave Remote Sensing of Snowmelt and Freeze/Thaw in the Kuparuk Basin, Alaska, using Calibrated Enhanced-Resolution Brightness Temperature (CETB) from SSMI/S and SMAP

VICKI JAGDEO[1], JOAN M. RAMAGE[1], MAHBOUBEH BOUESHAGH[1], AND MARY J. BRODZIK[2]

ABSTRACT

The relationship between snow and permafrost on Arctic terrain has significant implications for water storage, flooding hazards, and infrastructure vulnerability. Studies show that earlier snowmelt and high runoff cause increased permafrost thaw and earlier freeze/thaw cycles, which can be especially problematic for communities in Arctic regions facing disproportionate warming from climate change. To analyze permafrost vulnerability due to snow dynamics, we compared snowmelt and soil moisture in the Kuparuk River Basin in Northern Alaska from 2010-2019. We used Calibrated, Enhanced-resolution Brightness Temperatures (CETB) from two passive microwave sensors, Special Sensor Microwave Imager/Sounder (SSMIS: F18) 37 GHz vertically polarized channel and Soil Moisture Active Passive (SMAP) at 1.41 GHz vertically and horizontally polarized channels. Snowmelt onset timing and duration are derived from SSMI/S brightness temperature (Tb) and diurnal amplitude variations (DAV) threshold exceedance which is sensitive to changing levels of liquid water content. Additional pixel level standard deviation statistics were used to understand dynamics and finer spatial differences. Soil thaw and refreeze dynamics are captured using the Normalized Polarization Ratio successfully utilized with SMAP data3, along with SMAP Level-3 L-band radiometer soil products. The end of high DAV corresponds with snow saturation and snow depletion. It appears that soil thaw is closely aligned in time with high DAV and continues to fluctuate above seasonal freeze thresholds until snow accumulation. We investigate the spatial patterns of multiple factors to test whether early snow melt, longer melt refreeze, early snow saturation and disappearance increase permafrost vulnerability at different elevations and environments.

[1] Department of Earth & Environmental Sciences, Lehigh University, Bethlehem, PA, USA
[2] NSIDC/CIRES, University of Colorado Boulder, Boulder, CO, USA
Corresponding author: vsj221@lehigh.edu

Comparing Active and Passive Observations of Snowmelt Refreeze in the Sierra Nevada

GEORGE DUFFY[1], SAM TUTTLE[1], CARRIE VUYOVICH[2], ELIAS DEEB[3], AND ANGELA RIENZO[1]

ABSTRACT

The diurnal snowmelt refreeze cycle is the earliest detectable signal of melting snowpack, and therefore a valuable forecast tool for freshwater resources. There are two methods for detecting snowmelt refreeze cycles from space. Synthetic aperture radar (SAR) from the sentinel 1A and 1B satellites can provide occasional diurnal observations in certain regions of the planet. Brightness temperature measurements at 37V GHz from satellites such as the Special Sensor Microwave Imager / Sounder (SSMIS) fleet can be used to retrieve snowmelt refreeze signals with broader coverage and higher repeat frequency, but at a lower resolution. We have proposed a way to increase the resolution of near-daily brightness temperature snowmelt refreeze cycle observations by combining the two satellite methods when there are simultaneous observations. Towards this end, we provide the first direct comparison of SAR and brightness temperature snowmelt refreeze in the Sierra Nevada and the Yosemite area.

[1] Department of Earth and Environmental Sciences, Syracuse University, Syracuse, NY, USA
[2] NASA Goddard Space Flight Center, Greenbelt, MD, USA
[3] Cold Regions Research and Engineering Laboratory, Hanover, NH, USA
Corresponding author: george.a.duffy@gmail.com

Altimetric Ku-band Radar Observations of Snow on Sea Ice Simulated with SMRT

JULIEN MELOCHE[1], MELODY SANDELLS[2], HENNING LÖWE[3], NICK RUTTER[4], GHISLAIN PICARD[5], RANDY SCHARIEN[6], JOHN YACKEL[7], AND ALEXANDRE LANGLOIS[8]

ABSTRACT

Sea ice thickness is essential for climate studies and numerical weather prediction. Radar altimetry has provided sea ice thickness measurement since the launch of ERS-1 and currently through CryoSat-2, but uncertainty in the scattering horizon used to retrieve sea ice thickness arises from interactions between the emitted signal and snow cover on the ice surface. Therefore, modelling the scattering of the electromagnetic waves with the snowpack and ice is necessary to retrieve the sea ice thickness accurately. The Snow Microwave Radiative Transfer (SMRT) model was used to simulate the altimeter waveform echo from the snow-covered sea ice. A new field campaign was conducted in the Deese Strait near Cambridge Bay, Nunavut, Canada in April 2022. Measurements included microstructure from x-ray tomography and roughness measurements from structure from motion. These data and datasets from ground campaigns linked to CryoVex 2017 near Alert, Canada, and Operation Ice Bridge 2016 in Eureka Sound, Canada, were used to parameterize SMRT. Evaluation of SMRT in altimeter mode was performed against CryoSat-2 waveform data in pseudo-Low Resolution Mode. Simulated and observed waveforms showed good agreement, although it was necessary to retrieve ice surface roughness. Retrieved ice surface roughness in Cambridge Bay was 1.7 mm, which was close to the observed value of 1.4 mm for flat ice. A pseudo low resolution mode correction factor was required in the radar equation to capture the pulse peakiness in Cambridge Bay. Simulation of Ku-band and Ka-band backscatter in preparation for the ESA CRISTAL mission demonstrated the dominance of scattering from the snow-sea ice interface at Ku-band, with some surface scattering particularly for first year ice. However, at the Ka-band, while the scattering from the snow surface dominates, the snow-sea ice return is non-negligible for first year ice, and volume scattering is more prevalent for multiyear ice. This is the first study to consider scattering within the snow and demonstrate the origin of CryoSat-2 signals. SMRT can be used to develop a physical retracker algorithm to retrieve snow depth and sea ice thickness for radar altimeter missions.

---

[1] Département de géomatique appliquée, Université de Sherbrooke, Sherbrooke, Canada.

[2] Department of Geography and Environmental Sciences, Northumbria University, Newcastle-Upon-Tyne, UK.

[3] WSL Institute for Snow and Avalanche Research SLF, Davos, Switzerland.

[4] Department of Geography and Environmental Sciences, Northumbria University, Newcastle-Upon-Tyne, UK.

[5] IGE, Université Grenoble Alpes, Grenoble, France.

[6] Department of Geography, University of Victoria, Victoria, Canada.

[7] Department of Geography, University of Calgary, Calgary, Canada.

[8] Département de géomatique appliquée, Université de Sherbrooke, Sherbrooke, Canada.

Oral Session #2

Session Chairs: Zachary Fair & Craig Smith

Examination of Microwave Backscatter of Freshwater Lake Ice using Polarimetric Decomposition

CONNOR MCRAE-PHARO[1] AND GRANT GUNN[1]

ABSTRACT

Polarimetric decomposition is an invaluable technique for analyzing microwave signals reflected off freshwater ice features. In this study, we utilize Radarsat 2 quad-pol captures to investigate backscatter from two lakes: Old Crow Lake in the Yukon and Noel Lake in the Northwest Territories, each representing a contrast in morphometries (shallow and deep, respectively). The contrast in depth is designed to present two unique ice compositions, accounting for the presence and absence of tubular bubbles at the ice-water interface. Polarimetric decomposition allow for the identification of the dominant scattering mechanisms that occur for ice with and without tubular bubbles. Results showed that the ice was characterized by dominant single-bounce scatter, regardless of morphometry, followed by orders of magnitude weaker volume-scattering signal and a slight to nonexistent double-bounce signal. These findings have important implications for remote sensing applications, such as opening the prospect of freshwater ice thickness retrieval as interferometry performs with higher confidence with single bounce interactions. Overall, the application of polarimetric decomposition of freshwater ice backscatter shows promise as a valuable tool for understanding the relative contribution of the physical properties of the ice to radiative transfer and can be used to improve our current methods of ice and climate monitoring in the cryosphere.

[1] Department of Geography and Environmental Management, University of Waterloo, Waterloo, ON, Canada

Corresponding author: cmcraeph@uwaterloo.ca

Remote Sensing and Cloud Computing: Determining Lake Ice Phenology using Google Earth Engine and Sentinel-1 SAR Imagery

BRENDAN WARK[1] AND GRANT GUNN[1]

ABSTRACT

Monitoring lake ice phenology is crucial for climate modelling and managing the logistics of daily life in remote northern communities. Determining lake ice phenology using *in situ* measurements is costly and time-consuming, and standard remote sensing techniques have substantial processing and storage requirements. This project blends remote sensing and cloud computing to make large-scale lake ice phenology monitoring fast and accessible with minor hardware requirements.

The project runs a dynamic threshold variability method (DVTM) algorithm on the Google Earth Engine (GEE) processing environment. The DVTM requires a single shapefile of up to 10,000 lakes as input, then collects all Sentinel-1 C-band synthetic aperture radar (SAR) acquisitions covering the lakes a given winter season. It then creates a range of thresholds based on how much the backscatter values within the lakes change between Sentinel-1 acquisitions. Whenever the range of thresholds is crossed, the algorithm flags that date as melt or freeze onset for that specific lake.

Algorithm validation has been conducted on three study areas with 400 lakes each and has been focused on the melt onset portion of the algorithm. All three study areas are within the Alaska North Slope. Based on results from the Canadian Lake Ice Model (CLIMo), temperature records from weather stations, ERA5 reanalysis data, and qualitative analysis of Sentinel-1 SAR imagery, the algorithm can accurately determine the melt onset date of 80% of lakes.

[1] Department of Geography and Environmental Management, University of Waterloo, Waterloo, ON, Canada

Corresponding author: bwwark@uwaterloo.ca

Monitoring Lake Ice Thickness Changes using Interferometric SAR

JAKE E. FERGUSON[1] AND GRANT E. GUNN[1]

ABSTRACT

Freshwater lake ice thickness is an important observable, both broadly relevant to human activities and a key driver of physical, chemical, and ecological processes in cold regions. In northern regions, where lakes contribute up to 40% of surface area, lake ice data is an especially important input into climatological and meteorological models. The remote nature and sheer number of Canadian lakes – over 900,000 greater than 10 hectares – makes satellite-based observation the only practical comprehensive monitoring method for many systems. Active microwave-band instruments are particularly appealing due to their ability to selectively penetrate cloud and snow cover by modulating the wavelength used; Differential Interferometric Synthetic Aperture Radar (DInSAR) has been proposed as a tool for monitoring ice thickness for some time, but research has been stymied by difficulty in acquiring quality image pairs and temporal coherence loss between pairs.

RADARSAT Constellation Mission (RCM) offers substantially improved revisit times (4-day); we evaluate its potential to overcome these limitations and extract lake ice thickness interferometrically. Interferograms are derived from a set of 5-m resolution compact-polarization images of Kluane Lake and Aishihik Lake, Canada. Topographic phase contributions are simulated using ArcticDEM and removed: resulting ice thickness growth measurements are then compared to thermodynamically simulated values (CLIMo using ERA5 reanalysis as input) and *in situ* measurements from the Kluane Lake Research Station where possible. Mean DInSAR-derived thickness growth shows generally good agreement with these datasets. Phase unwrapping can be complicated by the formation of long, thin zones of low coherence associated with ridging or cracking. These results support the potential use of DInSAR as a tool for high-resolution lake ice thickness monitoring.

---

[1] Department of Geography and Environmental Management, University of Waterloo, Waterloo, ON, Canada

Corresponding author: j5fergus@uwaterloo.ca

Snow Depth Mapping on Canada's Sub-Arctic Lakes

ALICIA POUW[1], ALEX MACLEAN[1], AND HOMA KHEYROLLAH POUR[1]

ABSTRACT

Snow accumulation on lake ice influences ice growth and the timing of ice melt due to snow's highly insulative and reflective properties. Therefore, the lake ice thickness is expected to vary spatially over the lake due to heterogeneous snow accumulation from wind redistribution and snowpack metamorphism. Although lake ice models perform well in simulating ice thickness, they are very sensitive to the value of snow depth and density. However, mapping snow depth spatially over the entire lake has presented challenges partially due to poorly known lake ice surface elevation and snow variability. Current snow depth and ice surface elevation observations are sparse and mostly restricted to point measurements. This study used a multi-sensor approach utilizing Ground Penetrating Radar (GPR) and Remotely Piloted Aircraft System (RPAS) acquisitions collected over Canadian sub-Arctic lakes to map the distributed snow depth during the 2021 to 2022 winter season. The resulting one-meter spatial resolution snow depth maps were compared to *in situ* observations, showing a relative error of 33% in December 2021 (RMSE = 4.15 cm, Bias = -1.52 cm), and 14% in March 2022 (RMSE = 5.87 cm, Bias = 4.64 cm). Using this multi-sensor approach, we have developed a framework for mapping the snow depth distributed on lakes. Simultaneously collecting ice thickness observations furthers understanding of the spatial relation between snow depth, ice thickness, and ice surface elevation. The findings of this research can lead to an improved understanding of snow and lake ice interactions, which is essential for northern communities' safety and well-being.

[1] Department of Geography and Environmental Studies, Wilfrid Laurier University, Waterloo, ON, Canada
Corresponding author: apouw@wlu.ca

Oral Session #3

Session Chairs: Fraser King & Vicki Jagdeo

It's all about Timing: Exploring the Relationship between Snowmelt and Caribou (*Rangifer tarandus*) Migration in the Northwest Territories of Canada

MARIAH MATIAS[1], JOAN RAMAGE[1], AND MARY BRODZIK[2]

ABSTRACT

Recent studies suggest that snow characteristics may be primary drivers of migration, largely due to caribou's high level of mobility and their dependence on landscape conditions for locomotion. To investigate whether and how snow characteristics such as melt/refreeze status and the presence of ice are related to caribou movement, we used GPS (Global Positioning System) tracking collar data provided by the Government of the Northwest Territories' Department of Environment and Natural Resources to identify individual animal location and migration patterns, with a focus on female animals in the Bathurst herd. We analyzed 117 individual female caribou with more than 30,000 observations between 2007 and 2016 from the Bathurst herd in the Northwest Territories of Canada. We used a hierarchical model to estimate the beginning, duration, and end of spring migration and compared these statistics against snowpack characteristics (i.e., the timing of melt onset and melt/refreeze cycles) which we derived from 37 GHz vertically polarized (37V GHz) Calibrated, Enhanced Resolution Brightness Temperatures (CETB) at 3.125 km resolution. We found that the start and end of the female herd's spring migration closely followed melt onset and migration duration was longer when melt onset occurred earlier, suggesting that melt onset events provide triggers for migration or favorable conditions that increase mobility. However, we did not find any significant relationships when we tested them at the individual level. A causal relationship between snowmelt timing and caribou migration would allow for anticipation of the herd's migratory behavior and potential shifts in herd ranges.

[1] Department of Earth & Environmental Sciences, Lehigh University, Bethlehem, PA, USA
[2] NSIDC/CIRES, University of Colorado Boulder, Boulder, CO, USA
Corresponding author: mtm220@lehigh.edu

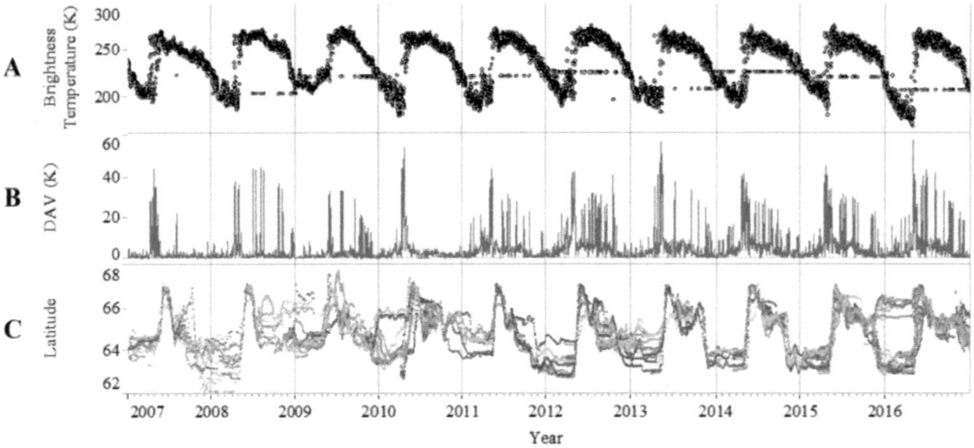

n = 20 n = 13 n = 15 n = 20 n = 18 n = 21 n = 14 n = 18 n = 32 n = 25

Figure 1. Time series of 3.125 km EASE-Grid pixels and female caribou latitudinal movement (northward migration) caribou departure locations between 2007 and 2016. The top two graphs are of brightness temperature (A) and diurnal amplitude variation (DAV) (B). The graph includes individual tracks of female caribou latitudinal movement through time (C). Individual female caribou are colored by unique collar ID. Collar ID legend has been omitted, as there are >100 unique IDs within this time span. Number of individuals per year is noted below each year's panel, where n = number of individuals.

SWE Retrieval Algorithm Advances using X- and Ku-band Radar

EDWARD KIM[1], FIROZ K. BORAH[2], LEUNG TSANG[2], AND DK KANG[3]

ABSTRACT

Snow water equivalent (SWE) retrieval via radar volume scattering at X and Ku bands (10 and 17 GHz) has been the basis of snow satellite concepts such as the European CoReH2O concept. In the decade since then, significant advances have been made in our understanding of the physics of the radar signature. And, critically, the retrieval approach has been validated through multiple airborne and tower-based field campaigns in several countries. This X/Ku algorithm now forms the basis of a new NASA snow mission concept that has put the snow community within grasp of achieving the long-desired goal of global SWE at high resolution for the first time.

At the core of the algorithm is a dense media radiative transfer (DMRT) model. A parameterization study demonstrated that the problem can be simplified to one of solving for just two unknowns—SWE and the microwave scattering albedo—removing the need for high-precision snowpack modeling. The retrieval solution is determined through a cost function minimization approach after accounting for surface scattering from the snow-soil interface, low vegetation, and the scattering albedo. We will show validation results using both tower and airborne data up to 750mm of SWE and discuss the options for obtaining the necessary input parameters.

[1] NASA Goddard Space Flight Center, Hydrological Sciences Lab, Greenbelt, MD, USA
[2] University of Michigan, EECS Department, Ann Arbor, MI, USA
[3] NOAA Weather Program Office, Silver Spring, MD, USA
Corresponding author: ed.kim@nasa.gov

Using ICESat-2 Altimetry to Derive Snow Depth over the Boreal Forests and Tundra of Alaska in Support of the SnowEx 2022/2023 Campaign

ZACHARY FAIR[1], CARRIE VUYOVICH[1], AND TOM NEUMANN[1]

ABSTRACT

The SnowEx 2022/2023 campaign was conceived to address questions about snow water equivalent (SWE), snow depth, and albedo in boreal forest and tundra environments. A series of field and airborne missions were conducted to accomplish these objectives through snow pit measurements, synthetic aperture radar (SAR), and lidar, among other methods. There has also been an emerging interest in using surface elevation measurements from the Ice, Clouds, and Land Elevation Satellite-2 (ICESat-2) mission for snow depth estimates. In this work, we present preliminary snow depth retrievals as derived from a combination of ICESat-2 data and snow-off digital elevation models (DEMs). We compare the ICESat-2 results to snow depths gathered from airborne lidar in March and October 2022 over five field sites in Alaska (Bonanza Creek, Caribou Creek, Creamer's Field, Toolik Station, and the Arctic Coastal Plain). Our results show general agreement between ICESat-2 and lidar-derived snow depths, particularly when the terrain is flat, and vegetation is accounted for. The mean bias in ICESat-2 snow depth relative to airborne lidar is -0.08 m We will expand upon these results in future work, using comparisons with field data, SAR, and stereographic imagery.

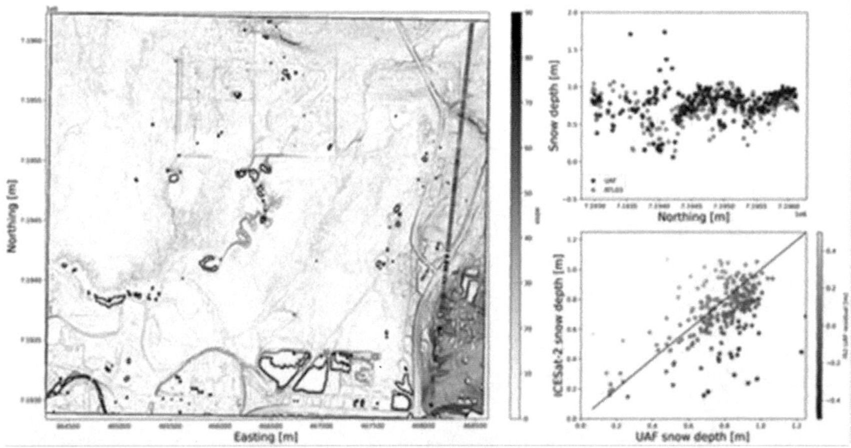

Figure 1. (Left) A slope map of Creamer's Field in Fairbanks, AK, derived using airborne lidar. The line is an ICESat-2 track that passed the field site on 21 March 2022, with the colors representing snow depth residuals between ICESat-2 and airborne lidar. (Top right) Along-track snow depth estimates for the ICESat-2 ATL03 product (green) and the UAF airborne lidar (purple). (Bottom right) Scatter plot of snow depths, with points colored by the snow depth residuals.

[1] NASA Goddard Space Flight Center, Greenbelt, MD, USA
Corresponding author: zachary.fair@nasa.gov

Verification and Analysis of the NOAA/NWS Baltimore/Washington Weather Forecast Office Winter Storm Threat Experimental Product

CONNOR BELAK[1] AND BRIAN LASORSA[1]

ABSTRACT

Information about winter storms occurring in the medium range (days 3-7) is in high demand by emergency managers, transportation planners, the general public, and broadcast/print media. The NOAA/National Weather Service Baltimore/Washington Weather Forecast Office (WFO LWX) developed and implemented an experimental product in 2016 that assesses and depicts both forecaster confidence and potential impact for winter storms in the medium forecast range. This publicly available experimental product is called the Days 3-7 Winter Storm Threat (WST). The WFO LWX Days 3-7 WST experimental product is designed for easy interpretation by general users, while also displaying uncertainty information for decision makers.

The WST product integrates data from the National Blend of Models (NBM), the National Centers for Environmental Prediction (NCEP)/Weather Prediction Center (WPC), and local expertise from WFO LWX staff. This local expertise is derived from forecaster experience working with core customers, combined with knowledge of local winter climatology and meteorology, to determine the level of winter storm threat. The product integrates locally known impacts and forecaster confidence to develop the winter storm threat to a region and is used by forecasters as a decision-making aid during impact-based decision support briefings.

This presentation will provide verification results of the WFO Days 3-7 WST from the winters of 2020-2021 through 2022-2023, analyze performance, and highlight several impactful events. Emphasis on this presentation will be the performance of the WST during Winter 2022-2023, when the algorithms were changed to become based on the NBM. The results of this research will guide future improvements to the Winter Storm Threat experimental product.

[1] NOAA/NWS Baltimore/Washington Weather Forecast Office, Sterling, VA, USA
Corresponding author: connor.belak@noaa.gov

Characteristics of Extreme Daily Snowfall Events near Arctic Coastal Regions

ALEKSANDRA ELIAS CHEREQUE[1], PAUL KUSHNER[1], LAWRENCE MUDRYK[2], AND CHRIS DERKSEN[2]

ABSTRACT

Climate change is causing the Arctic to warm, which has implications for conditions across North America and Eurasia. Changes in the Arctic affect seasonal snowpack conditions, sea ice, and soil characteristics, which might imply a changing susceptibility of large regions to disasters, such as wildfire or flooding, as warming continues. We have completed preliminary work looking at seasonal snowpack formation and trends by focusing on heavy snowfall events occurring early in the season. These were diagnosed and examined using reanalysis meteorological data and snow datasets generated by the recently updated Brown Temperature Index Model. Extreme events cause significant changes to the snow cover and snow depth on daily time scales. Across several Arctic regions, we identified important controls on these events, including a promising link between open (ice-free) ocean conditions and increased coastal snowfall in Eurasia. The results we present are robust across all datasets studied.

[1] University of Toronto, Toronto, Ontario, Canada
[2] Environment and Climate Change Canada, Climate Research Division, Toronto, ON, Canada
Corresponding author: akaecl@gmail.com

Poster Session #2

Session Chair: Krystopher Chutko

Daily Continental Scale Snow Water Equivalent Data for North America

BIDHYANANDA YADAV[1], MICHAEL DURAND[1,2], JACK DECHOW[2], SUJAY V KUMAR[3], AND MELISSA WRZESIEN[3]

ABSTRACT

Existing continental scale snow water equivalent (SWE) datasets are inadequate to meet many of the societal and scientific needs, such as understanding the impact of warming climate on cold region water cycle. Here, we present the preliminary result of our efforts to model the continental scale daily SWE over North America using a methodology called Blender. Blender is similar in theory to SWE reconstruction and is formulated as a mass-and-energy constrained optimization problem. The general methodology merges remote sensing observation of snow cover fraction with model outputs from land simulation models. Blender utilizes the MODIS cloud gap filled snow covered area (MODIS-CGF) to constrain the optimal estimate of snowfall, SWE, air-temperature, and ground-heat flux generated by Noah-MP simulations. The compute routine is implemented in Julia using Julia for Mathematical Optimization and interior point line-search algorithms. In the current setup we use results from Snow Ensemble Uncertainty Project (SEUP) which used the NASA Land Interface Systems Framework (LISF) running Noah-MP model simulations forced by Modern Era Retrospective Analysis for Research and Applications (MERRA-2) data at a nominal spatial resolution of 0.05 degrees (~5 km). The 0.5 km resolution MODIS-CGF snow cover dataset is resampled to 0.05 degrees SEUP resolution for the final Blender run to generate a daily SWE map of North America. Our future goal is to generate a higher resolution (~1 km) daily SWE map of North America for a 20-year period, spanning water years 2002 and 2021.

[1] Byrd Polar and Climate Research Center, The Ohio State University, Columbus, OH, USA
[2] School of Earth Sciences, The Ohio State University, Columbus, OH, USA
[3] NASA Goddard Space Flight Center, Greenbelt, MD, USA
Corresponding author: yadav.111@osu.edu

The Application of Disdrometers and Present Weather Detectors to Improve the Automated Measurement of Solid Precipitation

CRAIG D. SMITH[1] AND AMBER ROSS[1]

ABSTRACT

There are many advantages for automating precipitation measurement including remote operation, cost efficiencies, and increased measurement frequency that may not be realized with manual observation practices. However, the unsupervised observation of precipitation often increases the level of uncertainty in regards to quality assurance. In addition, key precipitation parameters, such as precipitation type, are unobserved. Popular automated all-weather weighing gauges used in cold regions generally depend on user-defined signal processing or require the user to trust on-board proprietary filtering algorithms to determine interval precipitation amount from bucket weight differential. Either of these techniques can benefit from ancillary information derived from disdrometers or present weather detectors. Two disdrometers (one optical and one radar based) were installed at two Canadian precipitation supersites (the Bratt's Lake and Caribou Creek Solid Precipitation Inter-Comparison Experiment sites) as a proof-of-concept for Environment and Climate Change Canada's (ECCC) operational networks and to support ongoing instrument intercomparison efforts. Although no practical automated reference exists at these sites to assess the accuracies of the two disdrometers, an ensemble approach incorporating the Double Fence Automated Reference (DFAR) for solid precipitation measurement is used to show their capabilities for improving precipitation detection, removing false precipitation reports, and deducing precipitation type for calculating rain:snow ratios and applying wind-bias transfer function adjustments.

Figure 1. Two disdrometers installed at the ECCC Bratt's Lake precipitation intercomparison supersite in Southern Saskatchewan Canada. Left: OTT Parsivel[2] optical disdrometer with the DFAR in the background. Right: Lufft WS100 Doppler radar disdrometer.

[1] Climate Research Division, Environment and Climate Change Canada, Saskatoon, Canada
Corresponding author: craig.smith@ec.gc.ca

Detecting Snow in Western New York and Eastern California using Sentinel-1a SAR and VIS/NIR Snow-Cover Maps

DOROTHY K. HALL[1,2], NICOLO E. DIGIROLAMO[2,3], AND GEORGE A. RIGGS[2,3]

ABSTRACT

Knowledge of the amount of snow and timing of snowmelt allows reservoir managers to optimize use of water resources and to improve predictions of water availability. With the currently available spaceborne visible/near-infrared (VIS/NIR) and synthetic aperture radar (SAR) satellite data, it is possible to map snow cover and to delineate areas of wet snow under certain circumstances. A great advantage of using SAR data for snow mapping is that data acquisition is unaffected by non-precipitating clouds, in contrast to VIS/NIR derived data products that can only image the surface under clear skies. The objective of this work was to determine the utility of Sentinel-1a (S-1a) C-band SAR data when used in combination with available snow-map products for mapping snow in western New York state and in a mountainous area in eastern California (Sierra Nevada (SN) mts.). We selected an initial study period from 1 December 2022 through 31 January 2023 for this analysis. In both study areas, using a time series of S-1a C-band SAR backscatter images, there were no significant backscatter, σ^0, changes indicating the presence of snow cover even when the VIS/NIR satellite snow-cover maps showed snow cover. The presence of cropland and deciduous forests in combination with the thin (generally < ~0.1 m), wet snow cover likely prevented the C-band SAR from detecting snow in the western New York study area. In the SN mts. study area during this time period, an expected decline in σ^0 as the snowpack developed was not observed. Further analysis of the SN snowpack was conducted during active melting in May of 2023 revealing a drop in σ^0 of up to ~4 dB as the melting progressed. Though promising, we consider these results to be preliminary because more data during the SN melting period are needed. Backscatter from snow at X/Ku-band is up to 40 times stronger compared to C-band and thus shows a great deal more potential for mapping both dry and wet snow as compared to C-band.

INTRODUCTION

Knowledge of the amount of snow and timing of snowmelt allows managers to optimize use of water resources and to improve predictions of water availability. When a snowpack becomes wet, melting may be imminent. With spaceborne visible/near-infrared (VIS/NIR) and synthetic aperture radar (SAR) data, it is possible to map snow cover and to delineate areas of wet snow, especially when meteorological station data are also available.

The objective of this work was to determine the utility of Sentinel-1a (S-1a) C-band SAR data used in combination with available snow-map products for mapping snow in western New York

[1] Earth System Science Interdisciplinary Center / University of Maryland, College Park, MD, USA
[2] Cryospheric Sciences Lab / NASA / Goddard Space Flight Center, Greenbelt, MD, USA
[3] SSAI, Lanham, MD, USA
Corresponding author: dkhall1@umd.edu

state and in the Sierra Nevada (SN) Mountains in eastern California. S-1a C-band SAR data have been shown to be useful for mapping wet snow when used in conjunction with VIS/NIR snow maps (e.g., Nagler et al., 2016; Solberg et al., 2016). A great advantage of using SAR data for snow mapping is that acquisition of data is unaffected by non-precipitating clouds, in contrast with use of visible/near-infrared (VIS/NIR) derived data products that can only image the surface under clear-sky conditions. Thus, a determination of the utility of available SAR data for mapping snow is important because of the predominance of clouds over so many areas of North America when snow is on the ground. Initially, we selected a study period from 1 December 2022 through 31 January 2023 for this analysis, but then added a second study period, 1 April – 10 May 2023, for the SN study area to investigate the C-band SAR response during spring melting.

We used snow map data products derived from the MODerate-resolution Imaging Spectroradiometer (MODIS) instruments on the Terra and Aqua satellites and the Suomi National Polar-orbiting Partnership (*NPP*) satellite, along with backscatter, σ^0, maps from the European Space Agency's (ESA's) S-1a C-band SAR (ESA, 2023).

STUDY AREAS

Western New York State

The New York study area (Fig. 1) is located in an area of western New York state that is predominantly cloudy for much of the fall and winter seasons.

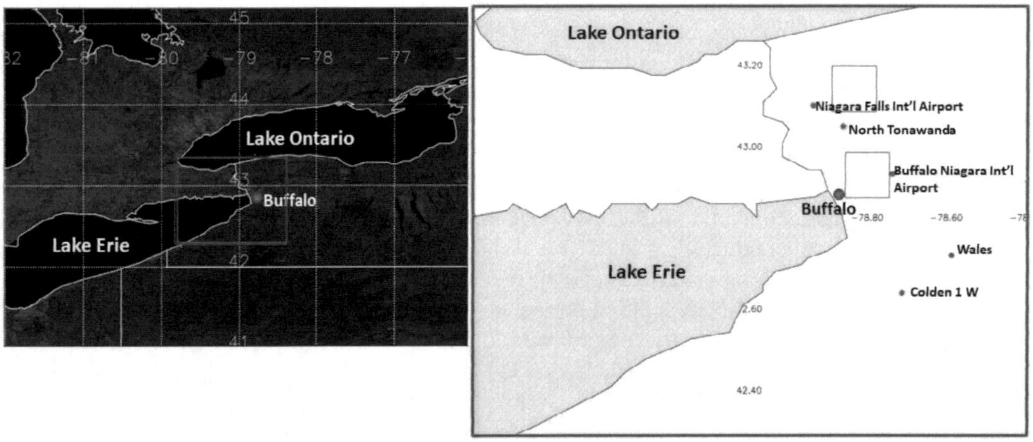

Figure 1. Left Panel - NASA Worldview base image showing the location of the study area in western New York state outlined in red (from the Blue Marble overlay based on an August 2004 shaded relief map (GIBS, 2004)). Right Panel– Zoom of study area. Locations of five meteorological stations are shown as red dots. The approximate location of the city of Buffalo, NY, is shown as a large red dot.

The New York study area is composed of croplands in the northern part and largely deciduous broadleaf forests in the southern part according to the 500-m resolution yearly MODIS-derived land-cover product, MODIS/Terra+Aqua Land Cover Type, MCD12Q1 (Friedl and Sulla-Menashe, 2019) (Fig. 2).

Figure 2. MODIS/Terra+Aqua Land Cover Type Yearly L3 map for 2021 MODIS/Terra+Aqua Land Cover Type, MCD12Q1 (Friedl and Sulla-Menashe, 2019) covering the study area shown in the right panel of Fig. 1. MCD12Q1 is based on the IGBP global vegetation classification scheme.

Sierra Nevada Mountains, California

The California study area (Fig. 3) is intermittently cloudy for much of the fall and winter seasons. It is composed of cropland, evergreen needleleaf forests and other according to the 500-m resolution yearly MODIS-derived land-cover product, MODIS/Terra+Aqua Land Cover Type, MCD12Q1 (Friedl and Sulla-Menashe, 2019) (Fig. 4). The highest elevation in the SN Mts. is ~4420 m.

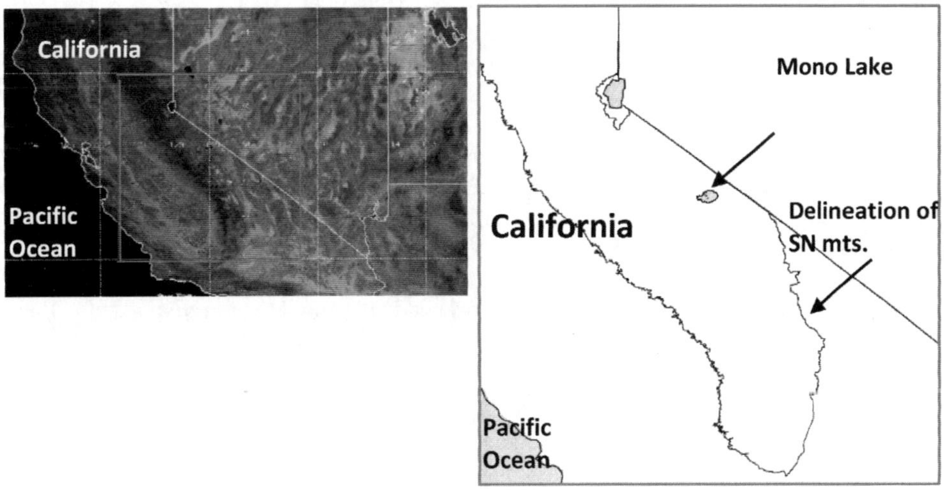

Figure 3. Left Panel - NASA Worldview base image showing the location of the study area in California outlined in red from the Blue Marble overlay based on an August 2004 shaded relief map (GIBS, 2004). Right Panel – Zoom of study area. The straight black lines delineate the California/Nevada border and the Sierra Nevada mts. are delineated using a shape file (see: https://sierranevada.ca.gov https://snc.maps.arcgis.com/home/webmap/viewer.html?webmap=66bb667660cb40b4a5b807192466f33d).

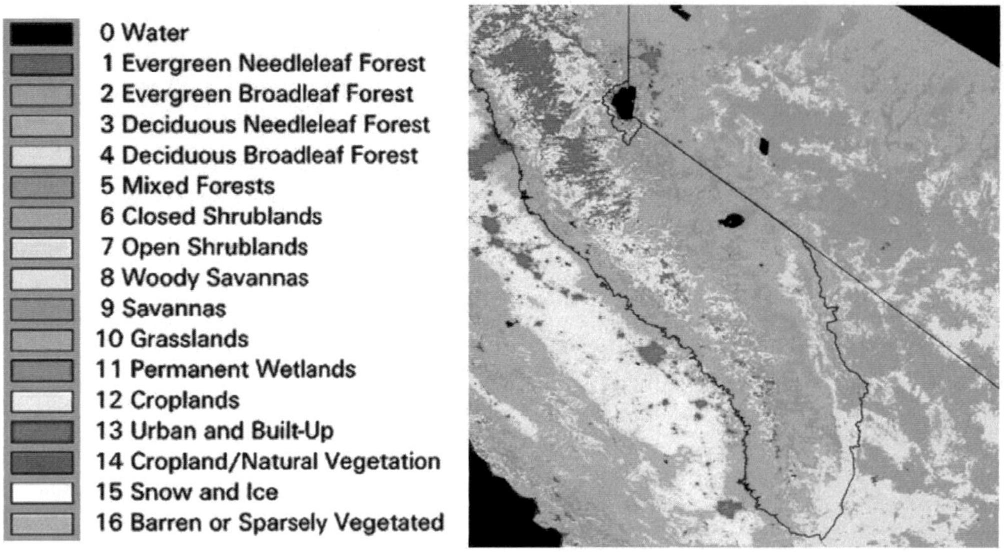

Figure 4. MODIS/Terra+Aqua Land Cover Type Yearly L3 map for 2021 MODIS/Terra+Aqua Land Cover Type, MCD12Q1 (Friedl and Sulla-Menashe, 2019) covering the study area shown in the right panel in Fig. 3. MCD12Q1 is based on the IGBP global vegetation classification scheme.

DATA

Snow-Cover Data Products

NASA standard snow-cover extent (SCE) data products, MOD10A1F and MYD10A1F, from the MODIS Terra and Aqua satellites, respectively (Hall and Riggs, 2020) and VNP10A1F from the Suomi National Polar Platform (SNPP) (Riggs et al., 2019) were used to map snow cover in the study areas. The MODIS and VIIRS data products were developed using a cloud-gap filling (CGF) algorithm, the details of which may be found elsewhere (Riggs et al. 2019 and Riggs & Hall, 2021). The accuracy of the CGF MODIS and VIIRS data products is compromised under persistently-cloudy conditions such as are found in the New York study area during late fall and winter seasons; for more accurate mapping, periodic views of the surface are needed to update the fully-automated snow decisions. Therefore, we also used the partially-automated Interactive Multisensor Snow and Ice Mapping System (IMS) 4-km resolution snow maps that are developed using a variety of satellite and station information and updated daily by trained meteorologists (Helfrich et al., 2018); IMS SCE maps are less affected by cloud-cover conditions on any given day than are the MODIS- and VIIRS-derived SCE maps.

There is a great deal of literature about the MODIS and IMS snow-cover products that will not be reviewed here. The MODIS SCE products have been available since February of 2000 and are produced in a completely automated computing environment. There have been several "collections," or versions of the MODIS SCE algorithms, each one providing improvements over the previous one. The current collection is Collection 6.1, or C6.1. The NASA VIIRS snow-cover products are modeled after the MODIS products, so details are not repeated here, but may be found in the user guides (Riggs et al., 2019; Riggs and Hall, 2021). We used the NASA VIIRS Collection 2 (C2) data products available from the National Snow and Ice Data Center (NSIDC, 2023).

Meteorological Station Data

All meteorological data used herein were obtained from NOAA (NCEI, 2023) or from the USDA (SNOTEL, 2022 - 2023). Five meteorological stations in New York were selected (Fig. 1 - right panel): Buffalo Niagara International Airport, Colden 1 W, Niagara Falls International Airport, North Tonawanda and Wales. For the SN study area, only the Virginia Lakes Ridge SNOTEL station in California was used; it is located to the north of Mono Lake in California (Fig. 5). We plotted the maximum (Tmax) and minimum (Tmin) daily air temperatures and snow depth during snowmelt in the months of April and May 2023. Using air temperature information, we estimated the state (wet vs dry) of the Sierra Nevada snowpack.

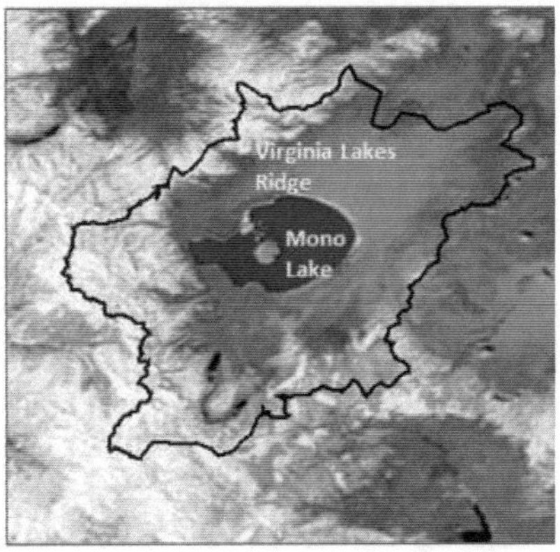

Figure 5. NASA Worldview image, 13 May 2023, showing Mono Lake in the central SN mts., eastern California. Mono Basin is outlined in black. The red dot to the north of the lake is the approximate location of the Virginia Lakes Ridge meteorological station.

Sentinel-1a

The S-1a satellite has a C-band SAR instrument on-board that was launched on 3 April 2014 with a repeat cycle of 12 days (ESA, 2023). We used the Interferometric Wide Swath (IW) mode, with VV+VH polarizations. IW is the nominal operation mode for land areas. The C-band SAR operates at a center frequency of 5.407 GHz, and has a swath width of 250 km and a nominal resolution of 5 m X 20 m. Details about the S-1a SAR may be found elsewhere (ESA, 2023).

For the New York study area, we used σ^0 maps derived from the C-band SAR acquired on 10 & 22 December 2022 and 3, 15 & 27 January 2023. The 10 December 2022 SAR σ^0 map was used as a snow-free reference map. For the California study area, we used σ^0 images from 8 December 2022 and 25 January 2023 as well as a snow-free reference image from 28 October 2022. Additionally, for the SN study area, we downloaded two S-1a SAR images acquired during active melting in the spring. The dates of the spring SAR images are: 8 & 13 May 2023.

Using S-1a C-band SAR to Identify Areas of Wet Snow

In earlier work (see Bernier and Fortin, 1998; Bernier et al., 1999; Baghdadi et al., 2000) it was reported that both wet and dry snow could be mapped successfully using C-band Radarsat data and even that snow-water equivalent (SWE) could be estimated (Bernier et al., 1999). However in recent literature, it has not been corroborated (to our knowledge) that C-band SAR data, for example from the S-1a satellite, is universally useful for determination of SWE, though Lievens et al. (2019) have reported success in mapping snow depth in deep snow in mountainous areas with S-1a C-band SAR data.

The combined use of data from optical and SAR sensors may enable the mapping of total snow area and monitoring of the temporal dynamics of snowmelt as described by Nagler et al. (2008 & 2016). Reduction of the backscattering coefficient, σ^0, in wet snow land surfaces compared to surfaces covered by dry snow or no snow is the basis for mapping wet snow areas with C-band SAR. A 3-dB threshold drop between non-snow-covered area or dry snow and wet snow has been noted in numerous works (e.g., Nagler et al., 2008; Lund et al., 2022; Solberg et al., 2016; Marin et al., 2020; and Manickam and Barros, 2020). Wet snow may be detected in a SAR image by comparing the σ^0 values with σ^0 from a reference image or images obtained during snow free or dry snow conditions (Nagler and Rott, 2000; Nagler et al., 2016). By using the average of two or more maps as a reference map, the classification accuracy can be improved (Nagler et al., 2008). Solberg et al. (2016) found a threshold of -3 dB for identifying wet snow using both C-band HH Radarsat (5.3 cm wavelength) and C-band VV ERS-SAR data. They used both SAR signatures and backscatter modeling to develop an algorithm for mapping wet snow in mountainous areas.

Marin et al. (2020) and Lund et al. (2022) compared S-1a backscatter with *in situ*-derived snow properties and snow model simulations. Marin et al. (2020) noted a decrease in C-band SAR σ^0 as soon as liquid water appeared in the snowpack. Lund et al. (2022) found thresholds of -2 to -3 dB to identify wet snow for co-polarized S-1a C-band SAR σ^0. To compare with the SAR-derived results, they used the SnowModel to provide an independent estimate of snowpack conditions. They also used >50 snow pit observations during the 2020 SnowEx experiment in Grand Mesa, Colorado, USA, for validation and concluded that C-band SAR is a reliable method to detect meltwater in a snowpack; (Lund et al., 2022).

For wet snow the radar signal at C-band frequencies is reflected and scattered at the surface and within the top few centimeters of the snowpack, whereas in the case of dry seasonal snow, a large part of the backscatter is contributed by the snow/ground interface (e.g., see Rott et al., 1992). For wet snow with liquid water content of 5% by volume, the penetration depth is only ~3 cm at C-band (Matzler, 1987; Nagler et al., 2016).

METHODOLOGY

For each study area, we compared the daily MODIS, VIIRS and IMS SCE maps for the study period from 1 December 2022 through 31 January 2023. Because of the persistent clouds in western New York during the study period, we used the IMS daily SCE to map the snow. For the eastern California/SN study area there were frequent views of the surface thus we were able to use the higher-resolution MODIS and VIIRS daily cloud-gap-filled snow maps to map the snow.

Next, we developed S-1a C-band σ^0 reference maps from scenes acquired when there was no snow on the ground as determined from MODIS, VIIRS and IMS SCE snow maps and NASA Worldview (2022-2023). For the western New York study area, an S-1a SAR scene from 10 December 2022 was used as the reference because we were able to confirm that there was no snow on the ground on that date. For the California/Sierra Nevada mts. study area, an S-1a-SAR scene from 28 October 2022 was used because there was no snow in the mountains visible on Worldview or the VIS/NIR snow products on that date.

We plotted the maximum and minimum daily air temperatures for each of the stations for the months of December 2022 and January 2023. Using air temperature data, we estimated the state (wet or frozen) of the snowpack surface. Because of the time of SAR image acquisition (~6PM local time), we used the maximum air temperature reported on that day as an indicator of the state of the snow surface. Using the above-mentioned methodology, we estimated the snow conditions on each date on which a SAR σ^0 map was available as seen in Tables 1 & 2.

Table 1. Determination of snow presence and state (wet or frozen) for each day on which an S-1a C-band SAR image was available for the western New York study area.

Date (and time in local time) of SAR image acquisition	Snow or no snow in scene	State of the snowpack at the time of SAR data acquisition
10 December 2022 (6:08 PM)	no snow	n/a
22 December 2022 (6:08 PM)	snow	wet
03 January 2023 (6:08 PM)	no snow	n/a
15 January 2023 (6:08 PM)	snow	frozen
27 January 2023 (6:08 PM)	snow	frozen

Table 2. Determination of snow presence and state (frozen or wet) of S-1a C-band SAR images for the Sierra Nevada Mts. study area. Other dates were downloaded but not used due to time constraints.

Date (and time in local time) of SAR image acquisition	Snow or no snow in scene	State of the snowpack at the time of SAR data acquisition
28 October 2022 (6:08 PM)	no snow	n/a
03 December 2022 (6:08 PM)	snow	unknown
20 January 2023 (6:08 PM)	snow	presumably frozen
08 May 2023 (6:08 PM)	snow	wet/melting

RESULTS

For each area, we show daily SCE for the study period from 1 December 2022 through 31 January 2023. Because of the persistently-cloudy conditions in western New York state during fall and winter, we used IMS SCE maps (Fig. 6). For the SN study area for which there were frequent views of the surface during the fall, winter and spring, we show VIIRS VNP10A1F SCE CGF maps (Fig. 7).

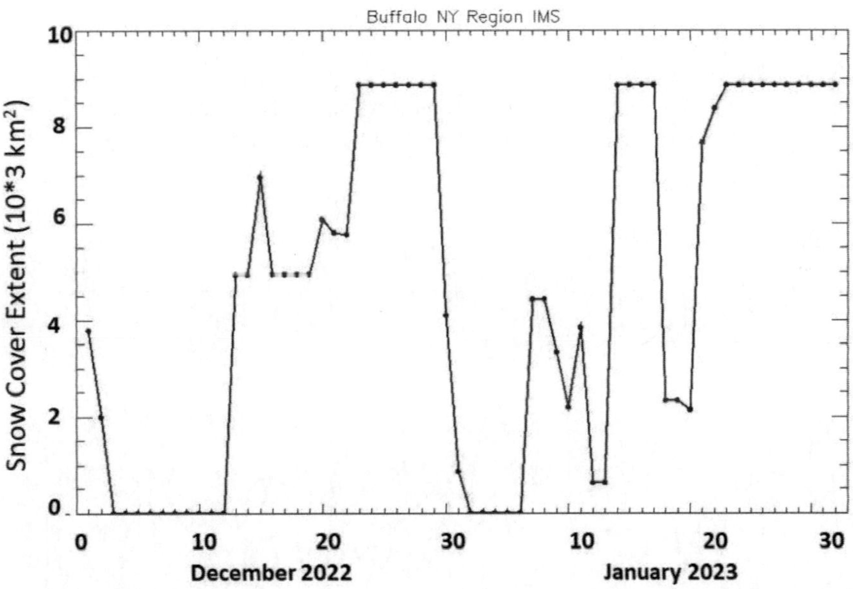

Figure 6. Daily SCE (in km^2) derived from the NOAA IMS daily snow maps in the western New York study area.

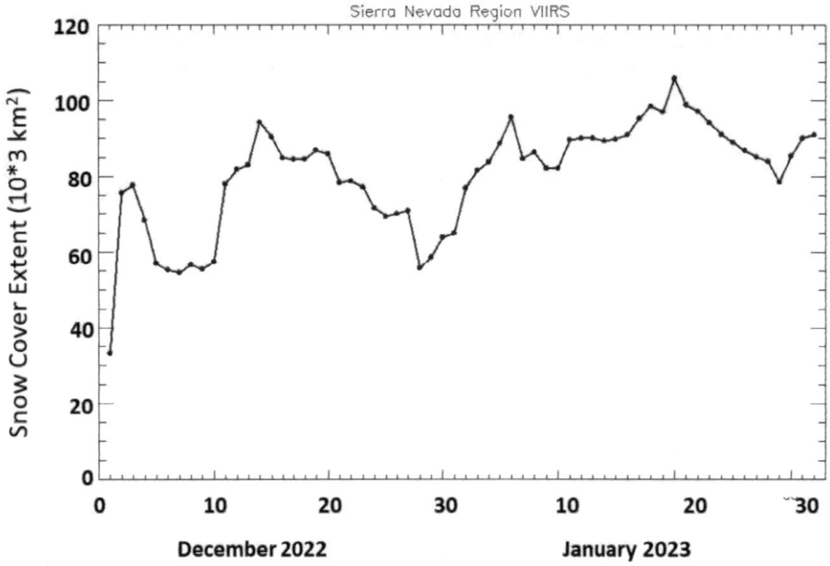

Figure 7. Daily SCE (in km^2) derived from NASA VIIRS VNP10A1F daily snow maps in the Sierra Nevada mts. study area.

Western New York study Area

From 1 December 2022 – 31 January 2023, we plotted Tmax, Tmin and snow depth from each of the five stations. An example from the North Tonawanda, NY, station is shown in Fig. 8.

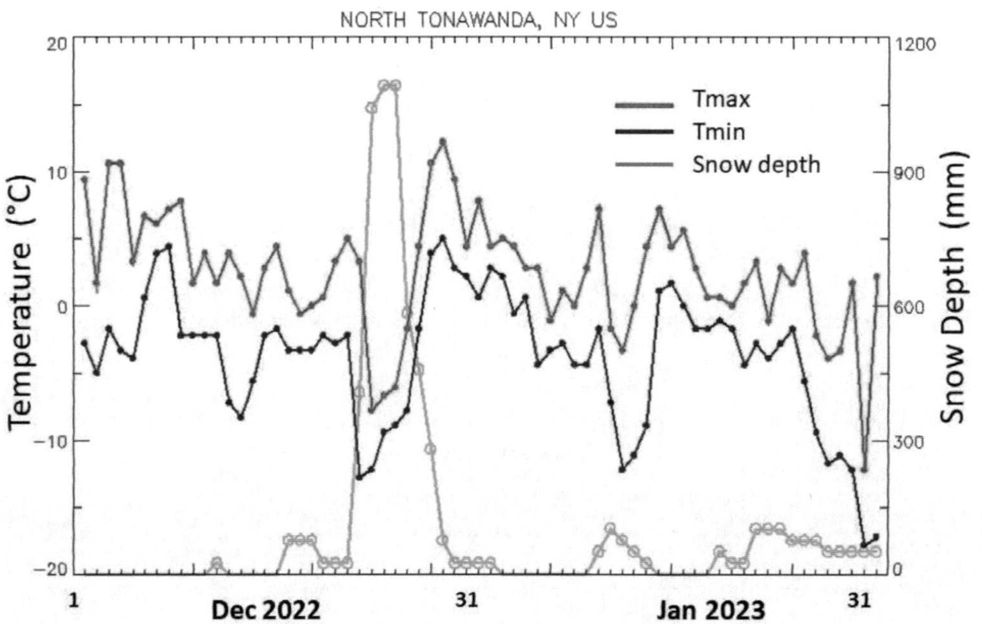

Figure 8. Daily Tmax and Tmin air temperatures (°C) and daily snow depths (mm) at North Tonawanda, NY, for the months of December 2022 and January 2023. Data were acquired from NCEI (2023).

In the land areas of the western New York study area, we did not find a significant difference in σ^0 on 22 December 2022 (when very thin snow was present (Fig. 8)) as compared to the reference image (10 Dec 2022 – not shown). Nor did we find a significant difference in the backscatter above and below the snowline on the 22 December 2022 cross-polarized SAR scene (Fig. 9a). The mean σ^0 of the land areas above (non-snow) and below (snow covered) the snowline on 22 December is -15.7515 dB, and -15.6401 dB, respectively. It is likely that the non-snow-covered land to the north of Lake Erie is wet (see Tmax in Fig. 8). It is difficult to distinguish between wet ground and wet snow using C-band SAR especially when the snow cover is thin (e.g., Nagler et al., 2016). Though the snow was deeper (up to >~ 1 m) between 23 and 31 December 2022 (Fig. 8), no S-1 C-band SAR images were available during this time period to assess the S-1a σ^0 in the deeper, wet snow.

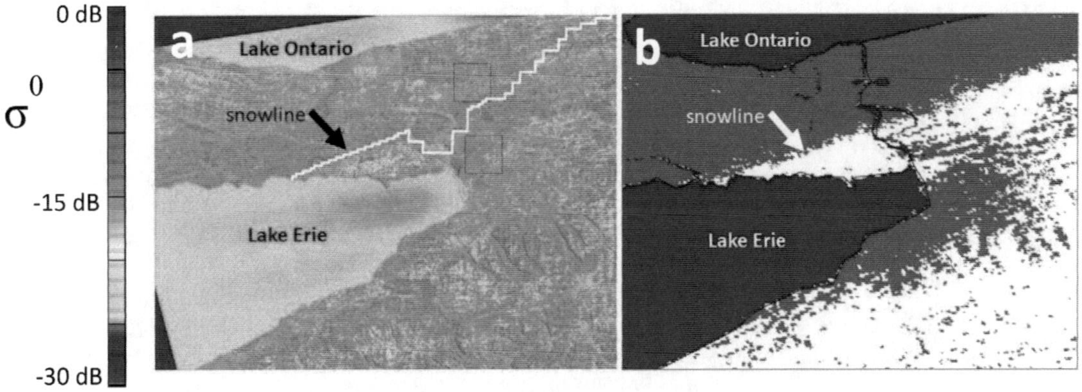

Figure 9. a) 22 December 2022 VH SAR σ^0 image. The snowline shown was derived from the 22 December 2022 IMS SCE map. b) 22 December 2022 VIIRS VNP10A1F snow map of the same area shown in the SAR image.

Sierra Nevada (SN) Mts. Area

For the Virginia Lakes Ridge SNOTEL station near Mono Lake in the SN mts., we plotted Tmax, Tmin and snow depth. Fig. 10 shows an example plot during the melt season, from 1 April – 10 May. During this time the snow was actively melting (as seen in the light blue line showing decline in snow depth).

The building snowpack in December 2022 and January 2023 did not produce significant differences in σ^0 as compared to the 28 October 2022 non-snow-covered reference image. We also developed a time series during active melting, 1 April – 10 May 2023. During active snowmelt in May 2023, the C-band SAR VV and VH σ^0 displayed a mean drop of up to 4.1 dB, within the land area of the Mono Lake Basin (Mono Basin) compared to a January 2023 σ^0 map, acquired when the snowpack surface was likely dry as determined from air temperature data (see Fig. 11a & b). (Note also the location and outline of Mono Basin in Fig. 5.) Though these preliminary results are inconclusive, they are consistent with available literature indicating that there is a 3 – 5 dB drop in σ^0 from dry to wet snow.

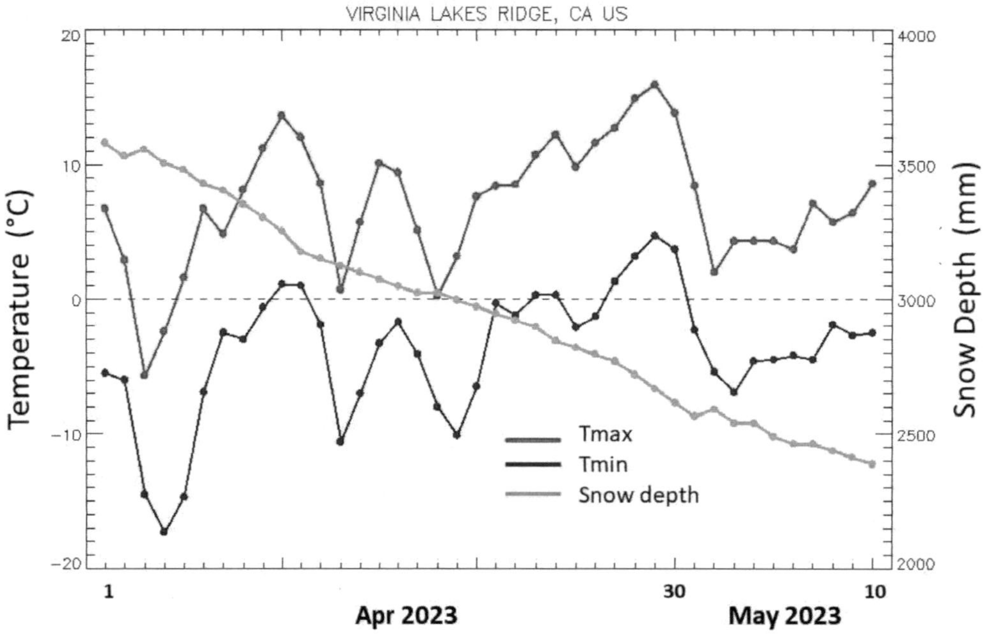

Figure 10. In the Sierra Nevada mts. study area (Fig. 3), Tmax at the Virginia Lakes Ridge station, northwest of Mono Lake, is consistently above 0°C from 5 April - 10 May 2023. Data acquired from SNOTEL (2022 - 2023).

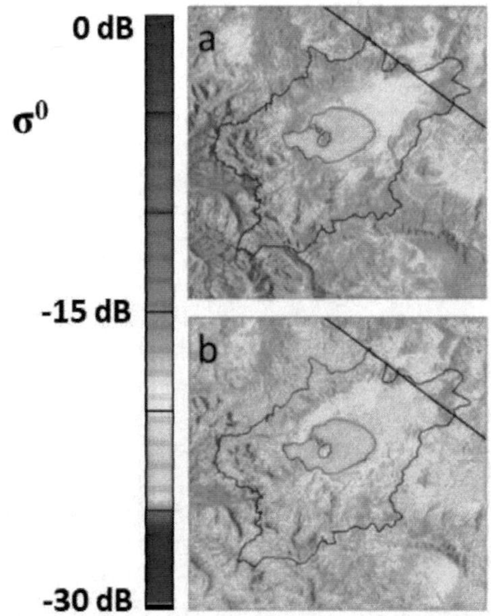

Figure 11. S-1a C-band SAR VH backscatter maps (ascending passes) focusing on Mono Basin. The thin black line delineates the basin. The thicker straight black line is the boundary line between California and Nevada. a) Acquired on 20 January 2023 when presumed dry snow covered the entire area; b) Acquired on 8 May 2023 during active snowmelt (see Fig. 10). Note the lower overall backscatter in b) as compared to a), possibly due to active snowmelt on 8 May.

DISCUSSION AND CONCLUSION

The 1 December 2022 – 31 January 2023 time series of MODIS and VIIRS cloud-gap-filled snow maps did not adequately map snow over the persistently-cloudy study area in western New York state because there were long stretches of time when the surface was not visible to optical sensors. The daily IMS snow maps are the most useful maps available for this study area during the fall and winter months due to the persistent clouds. In the SN study area over the 2022 – 2023 winter and spring, there were frequent views of the surface between storms and thus the MODIS and VIIRS cloud-gap-filled snow maps performed well for mapping snow there. In this paper, we show only results from IMS and VIIRS.

To map wet snow using C-band SAR data, a snow mask developed from optical data is needed (Nagler et al., 2008 and 2016; Solberg et al., 2016; Manickam & Barros, 2020). But even with an optically-derived snow mask, it may not be possible to map snow using C-band SAR data because of the similarity in backscatter, σ^0, between wet ground and wet snow. This is especially true when the snowpack is shallow and wet as was the case in western New York during our study period on the days for which S-1a C-band SAR data were available.

Specifically, we did not see a significant difference in σ^0 between snow-covered and non-snow-covered land areas in our study area in western New York from 1 December 2022 to 31 January 2023. Though we see a snowline on the 22 December scene on VIS/NIR imagery, the σ^0 does not change significantly above and below the snowline, even though many researchers have shown strong backscatter decreases (3 – 5 dB) between snow-free ground and wet snow in other study areas as discussed earlier. The reason that we are not seeing a σ^0 change is likely because the snow is wet and quite thin (mostly < 0.1 m) and the ground (mostly cropland) in the vicinity of the snow cover on 22 December 2022 is probably also wet based on reported Tmax temperatures. At C-band, backscatter from vegetation can significantly contribute to the total backscatter or even dominate the snow signature (Nghiem and Tsai, 2001).

To map both dry and wet snow using SAR data, higher microwave frequencies such as X/Ku-band have much more potential than C-band (e.g., Matzler, 1984). Backscatter from snow at Ku-band (~14 GHz) is 5.4 times stronger than at X-band (~9 GHz) and 40 times stronger compared to C-band (~5.3 GHz) (Nghiem and Tsai, 2001). Though there are many circumstances when the S-1a C-band SAR can map snow in conjunction with a VIS/NIR snow mask, it is not universally possible according to this preliminary work.

Acknowledgements. The authors thank HP Marshall/Boise State University, Thomas Nagler/Enveo IT GmbH, Austria, Ana Barros/University of Illinois and Surendar Manickam/Vellore Institute of Technology, India, for discussions about snow wetness and C-band SAR.

REFERENCES

Baghdadi N, Gauthier Y, Bernier M., Fortin JP. 2000. Potential and limitations of RADARSAT SAR data for wet snow monitoring. *IEEE Transactions on Geoscience and Remote Sensing,* **38**: 316-320.

Bernier M, Fortin JP. 1998. The Potential of Times Series of C-Band SAR Data to Monitor Dry and Shallow Snow Cover. *IEEE Transactions on Geoscience and Remote Sensing,* **36**: 226-243.

Bernier M, Fortin JP, Gauthier Y, Gauthier R, Roy R, Vincent P. 1999. Determination of snow water equivalent using RADARSAT SAR data in eastern Canada. *Hydrological Processes,* **13**: 3041-3051.

ESA. 2023. Sentinel-1 C-band Synthetic Aperture Radar (SAR), *https://www.earthdata.nasa.gov/sensors/sentinel-1-c-band-sar*

Friedl M, Sulla-Menashe D. 2019. MCD12Q1 MODIS/Terra+Aqua Land Cover Type Yearly L3 Global 500m SIN Grid V006 [Data set]. NASA EOSDIS Land Processes DAAC. Accessed 2023-03-16 from *https://doi.org/10.5067/MODIS/MCD12Q1.006*

GIBS (Global Imagery Browse Services) NASA Worldview. 2004. *https://worldview.earthdata.nasa.gov/*

Hall DK, Riggs GA. 2020. MODIS/Terra CGF Snow Cover Daily L3 Global 500m SIN Grid, Version 61 [dataset]. Boulder, Colorado USA. NASA National Snow and Ice Data Center Distributed Active Archive Center. *https://doi.org/10.5067/MODIS/MOD10A1F.061*

Helfrich S, Li M, Kongoli C, Nagdimunov L, Rodriguez E. 2018. IMS Algorithm 991 Theoretical Basis Document Version 2.4. 74.

Lievens H, Brangers I, Marshall H-P, Jonas T, Olefs M, De Lannoy G. 2022. Sentinel-1 snow depth retrieval at sub-kilometer resolution over the European Alps. *The Cryosphere*, **16:** 159–177.

Lund J, Forster RR, Deeb EJ, Liston GE, Skiles SM, Marshall H-P. 2022. Interpreting
Sentinel-1 SAR backscatter signals of snowpack surface melt/freeze, warming, and ripening through field measurements and physically-based SnowModel. *Remote Sensing,* **14**: 4002. *https://doi.org/10.3390/rs14164002*

Manickam S, Barros A. 2020. Parsing synthetic aperture radar measurements of snow in complex terrain: Scaling behaviour and sensitivity to snow wetness and landcover. *Remote Sensing*, **12:** 483. *https://doi:10.3390/rs12030483.*

Marin C, Bertoldi G, Premier V, Callegari M, Brida C, Hürkamp K, Tschiersch J, Zebisch M, Notarnicola C. 2020. Use of Sentinel-1 radar observations to evaluate snowmelt dynamics in alpine regions. *The Cryosphere,* **14**: 935–956.

Mätzler C, Schanda E. 1984. Snow mapping with active microwave sensors. *International Journal of Remote Sensing*, **5**: 409-422.

Mätzler C. 1987. Applications of the interaction of microwaves with the natural snow cover. *Remote Sensing Reviews*, **2**: 259-387.

NASA Worldview. (2022 - 2023). *https://worldview.earthdata.nasa.gov/*

NCEI. 2023. NOAA National Centers for Environmental Information, Daily Summaries Version 3.1.0, Global Historical Climate Network includes daily land surface observations from around the world [Dataset]. *https://www.ncei.noaa.gov/maps/daily-summaries/*

Nagler T, Rott H. 2000. Retrieval of wet snow by means of multitemporal SAR data. *IEEE Transactions on Geoscience and Remote Sensing,* **38**: 754-765.

Nagler T, Rott H, Malcher P, Muller F. 2007. Assimilation of meteorological and remote sensing data for snowmelt runoff forecasting. *Remote Sensing of Environment,* **112**: 1408-1420. *https://doi:10.1016/j.rse.2007.07.006*

Nagler T, Rott H, Ripper E, Bippus G, Hetzenecker M. 2016. Advancements for snowmelt monitoring by means of Sentinel-1 SAR. *Remote Sensing,* **8**: 348.

Nghiem SV, Tsai WY. 2001. Global snow cover monitoring with spaceborne Ku-band scatterometer. *IEEE Transactions on Geoscience and Remote Sensing*, **39**: 2118-2134.

NSIDC. 2023. *https://nsidc.org/data/viirs/data#top*

Riggs GA, Hall DK, Román MO. 2019. MODIS Snow Products Collection 6.1 User Guide Version 1. *http://modis-snow-ice.gsfc.nasa.gov/?c=userguides*

Riggs GA, Hall DK. 2021. NASA S-NPP VIIRS Snow Cover Products Collection 2 User Guide. *http://modis-snow-ice.gsfc.nasa.gov/?c=userguides*

Riggs GA, Hall DK, Román MO. 2019. VIIRS/NPP Snow Cover Daily L3 Global 375m SIN Grid, Version 1 [dataset]. Boulder, Colorado USA. NASA National Snow and Ice Data Center Distributed Active Archive Center. *https://doi.org/10.5067/VIIRS/VNP10A1.001*

Riggs GA, Hall DK. 2020. Continuity of MODIS and VIIRS snow cover extent data products for development of an Earth Science Data Record. *Remote Sensing*, **12**: 3781. *https://nsidc.org/sites/nsidc.org/files/technical-references/Riggs&Hall.2020.pdf*

Rott H, Davis RE, Dozier J. 1992. Polarimetric and multifrequency SAR signatures of wet snow. *Proceedings of the Internationl Geoscience and Remote Sensing Symposium* (26-29 May, 1992, Houston, Texas, USA): 1658-1660.

Sentinel-1 SAR. 2018. IEEE International Geoscience and Remote Sensing Symposium, Publisher: IEEE, 2018:8727-8730, *https://doi.10.1109/IGARSS.2018.8518203*, ISBN: 978-1-5386-7150-4.

SNOTEL. 2022 – 2023. USDA SNOTEL data for Virginia Lakes Ridge, California *https://wcc.sc.egov.usda.gov/nwcc/site?sitenum=846*

Solberg R, Rudjord Ø, Salberg AB, Due Ø, Trier GS, Diamandi A, Irimescu A. 2016. Single-and multi-sensor snow wetness mapping by Sentinel-1 and MODIS data. *73rd Eastern Snow Conference*, Columbus, OH.

Retrieval of Snowpack Density and Ice Grain Radius from Time-Domain Diffuse Optical Measurements

CONNOR HENLEY[1,2], JOSEPH HOLLMANN[2], COLIN MEYER[3], AND RAMESH RASKAR[1]

ABSTRACT

Lidar is widely used to map the geometry of the top surface of snowpacks. However, because snow is a collection of highly transparent and forward scattering ice grains, most photons in an incident laser pulse scatter many times beneath the surface before returning to the receiver. In this work we show that the measurable time delay caused by subsurface scattering events can be exploited to estimate volumetric properties of the snowpack - namely, snowpack density, ice grain radius, and impurity concentration.

We model the propagation of laser light within the snowpack using a photon diffusion model that was originally developed by the medical imaging community to model light transport in human tissue. We show that, for clean snow, the parameters of this model - the absorption coefficient, effective scattering coefficient, and mean speed of light - can be written solely as functions of grain radius and snowpack density if grains can be approximated as spherical.

We illuminate the snowpack surface at a single point using a pulsed laser source, and use a photon-counting SPAD detector to measure the time-dependent intensity of light that exits the snowpack a small distance (5-10 cm) from the laser spot. We fit a curve to these measurements that has the same form as the appropriate solution to the time-dependent photon diffusion equation, and then choose the grain size and density that can best reproduce the parameters of the fitted curve. We validate our method using measurements collected for a variety of snow types.

[1] MIT Media Lab, Massachusetts Institute of Technology, Cambridge, MA, USA
[2] The Charles Stark Draper Laboratory, Inc., Cambridge, MA, USA
[3] Thayer School of Engineering, Dartmouth College, Hanover, NH, USA
Corresponding author: co24401@mit.edu

MODIS-VIIRS Snow Cover Extent Data Product Continuity

GEORGE A. RIGGS[1] AND DOROTHY K. HALL[2]

ABSTRACT

NASA snow cover data products have been produced since 2000 from the satellite and instrument series of the Terra and Aqua Moderate-resolution Imaging Spectroradiometer (MODIS), Suomi-National Polar Program (SNPP) Visible Infrared Imaging Radiometer (VIIRS), Joint Polar Satellite Series (JPSS) from satellites JPSS-1 and JPSS-2. In 2000 there was a single daily NASA satellite snow cover extent (SCE) data product from MODIS and, in 2023, there are now five daily NASA satellite SCE data products. A 23-year record of SCE composed of observations from five similar instruments has great potential for study of snow cover climatology changes and trends. Though the satellites and instruments are similar there are differences in orbits and instrument characteristics that can affect the detection of snow cover in the automated algorithms that produce the products. An algorithm continuity approach, using the same algorithm for both sensors, was taken to maximize product continuity. We present evidence of daily continuity in the NASA SCE products, and continuity of seasonal changes of SCE trends among the products.

INTRODUCTION

For more than 20 years the Moderate-resolution Imaging Spectroradiometer (MODIS) instruments on the Terra and Aqua satellites have provided daily near global coverage of Earth. The Terra and Aqua missions have far surpassed their original designed five-year lifetime. NASA has initiated preparing end-of-life plans for the aging Terra and Aqua missions. Terra and Aqua are now in orbital drift but will continue to observe Earth until passivation of the satellites occurs probably in 2025. The snow cover extent (SCE) records from the MODIS instruments have been used extensively by the scientific community for research and applications. How will the 20+ year record of MODIS snow cover products be continued? The NASA VIIRS snow cover products will extend the SCE Earth Science Data Record (ESDR) that began with MODIS in 2000. A VIIRS instrument is onboard the Suomi-National Polar-orbiting Partnership (SNPP), the Joint Polar Satellite System series of satellites (JPSS) JPSS-1 (NOAA-20) and JPSS-2 (NOAA-21) satellites and is planned to fly on JPSS-3 and JPSS-4 satellites likely extending the SCE ESDR to about 2040.

Currently, five polar orbiting satellites acquiring daily observations of the Earth are used to generate daily snow cover products as shown in the timeline in Figure 1. The observation history and foreseeable future of the satellites is time scaled to show overlap of satellites.

[1] Science Systems and Applications, Inc., Lanham, MD, USA.
[2] Earth System Science Interdisciplinary Center/University of Maryland, College Park, MD, USA.
Corresponding author: george.riggs@ssaihq.com

Continuity is the situation where a data product continues through a change in data source in this case MODIS ends and the snow cover data products continue with VIIRS. Data continuity is a prerequisite for long-term monitoring of change and identifying trends (Román et al., 2023). The objective of this analysis is to evaluate continuity among the MODIS and VIIRS snow cover data products to determine their suitability for providing an SCE data record suitable for research on snow cover climatology.

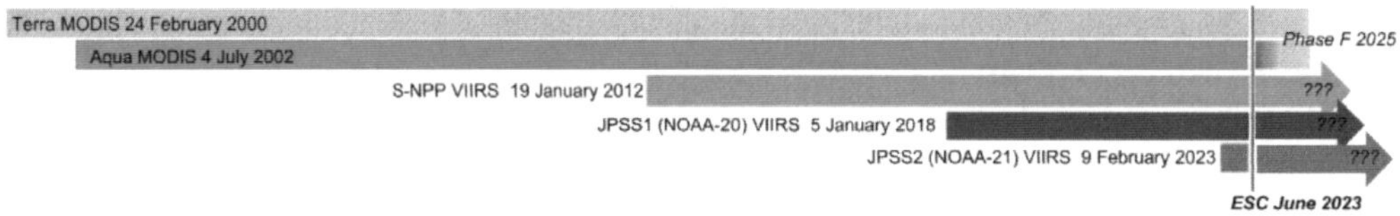

Figure 1. Timeline of Terra, Aqua, SNPP, JPSS-1, and JPSS-2 satellites. Time scaled to show overlap of satellites.

DATA PRODUCTS

Continuity among the MODIS Terra, Aqua, and VIIRS SNPP and JPSS-1 daily and daily cloud-gap-filled (CGF) products has been investigated. The data products are listed in Table 1. The MODIS and VIIRS CGF products provide a daily "cloud free" view of SCE. Data products are available from the National Snow and Ice Center (NSIDC) Distributed Active Archive Center (DAAC) which provides user support for the products, user guides and software tools. Details of the MODIS data products are described in the user guide (Riggs et al., 2019a) and the VIIRS data products are described in the user guides (Riggs et al., 2019b, and Riggs and Hall, 2021). The JPSS-2 data products (Table 1) are produced in the Land Science Investigator-led Processing System (LSIPS) and will be released to the community in the near future.

Table 1. Daily MODIS and VIIRS snow cover data products.

Instrument/ Satellite	ESDT Daily	ESDT Daily cloud-gap-filled	Resolution	Collection	Format
		MODIS			
Terra	MOD10A1	MOD10A1F	500 m	C61	HDF4EOS
Aqua	MYD10A1	MYD10A1F	500 m	C61	HDF4EOS
		VIIRS			
SNPP	VNP10A1	VNP10A1F	375 m	C2	HDF5EOS
JPSS-1 (NOAA-21)	VJ110A1	VJ110A1F	375 m	C2	HDF5EOS
JPSS-2 (NOAA-22)	VJ210A1	VJ210A1F	375 m	Not yet released	HDF5EOS

SNOW COVER EXTENT COMPARISONS

The MODIS and VIIRS data products contain several data layers, described in the user guides. We used the NDSI_Snow_Cover data layer from the daily products, and the NDSI_Snow_Cover_CGF data layer from the CGF products. Comparisons of the snow cover products were made for three regions with differing characteristics and different time periods to evaluate product continuity in different snow cover situations. The snow cover products were compared either by NDSI_Snow_Cover or NDSI_Snow_Cover_CGF data layer values or by binary SCE. SCE was determined by converting the NDSI_Snow_Cover or

NDSI_Snow_Cover_CGF data layers to binary snow cover by defining NDSI values of 10-100 as snow cover. NDSI values of 0-9 are defined as snow free land. Binary SCE is used for comparisons because convolved factors make it very difficult to compare NDSI_Snow_Cover values directly as discussed in Riggs and Hall (2020). Snow cover extent is the total area of snow cover observed on a day, calculated as the number of cells with snow cover then multiplied by cell area. The JPSS-2 products (Table 1) were not included in the analysis because they are not currently available to the community and not available for dates of some of the comparisons that have been done.

Pacific Northwest, USA Region

The Terra, Aqua, SNPP and JPSS-1 daily SCE and daily CGF products (Table 1) were compared to evaluate continuity among the four satellites. The objective of this comparison was to evaluate continuity among the products over a few consecutive days and evaluate whether the single day product that includes cloud cover or the "cloud free" CGF snow cover product provides the best product for monitoring snow cover. The SCE products for a single tile (tile h09v04) covering the Pacific Northwest region of the USA, from 1 – 11 March 2023 were compared.

The daily SCE maps (Table 1, *10A1) for 1–3 March 2023, which had the least cloud cover in the period are shown in Figure 2 to illustrate typical similarities and differences among the SCE maps. SCE is very similar among the MODIS and VIIRS products each day and is spatial consistent day-to-day (Figure 2). Cloud cover may or may not be the same among the products because of differences in times of acquisition, clouds may change in location, formation, and extent thus obscuring and changing the extent of snow observed (Figure 2). The acquisition time difference is greatest between Terra with a morning local crossing time and the afternoon local crossing times of Aqua, SNPP and JPSS-1. The Aqua, SNPP, and JPSS-1 crossing times are less than about 30 min apart. Also, the MODIS and VIIRS cloud mask algorithms are different in cloud detection tests applied because of spectral band differences between MODIS and VIIRS. So even if cloud cover is the same in two different products, for example, a result may be different if the cloud masking differs.

Cloud cover makes it difficult to compare SCE maps among the products because clouds obscure the surface and clouds can change between satellite acquisition times. In the daily product SCE maps (Figure 2) cloud cover in this period is consistent among the products however on 3 March the MYD10A1 has a greater cloud cover extent in the southwest sector of the tile than does the Terra MOD10A1. Perusal of Worldview (worldview.earthdata.nasa.gov) imagery of Terra and Aqua reveals that there was an increase in cloud cover in that region by the time of the Aqua acquisition. Clouds obscured some mountain cloud cover however that cloud cover changed in formation and location between the Aqua and SNPP acquisitions and some of the mountains snow cover that was obscured in Aqua was visible in the SNPP acquisition, thus explaining the SCE differences between Aqua and SNPP (Figure 2). That is an example of the level of investigation that sometimes needs to be done to evaluate daily differences in the daily SCE products that are caused by cloud cover.

The spatial extent of snow cover and clouds in the products for 1–11 March 2021 is plotted in Figure 3A. The extent of snow cover is similar among the products on most days but some days e.g., 2 March, the MYD10A1 (MODIS Aqua) SCE is notably less than the other products. On days 4–7 March MODIS SCE is similar but is notably less than SCE in the VIIRS products. Spatial extent of cloud cover tends to be more similar among the products than is SCE as shown in Figure 3B.

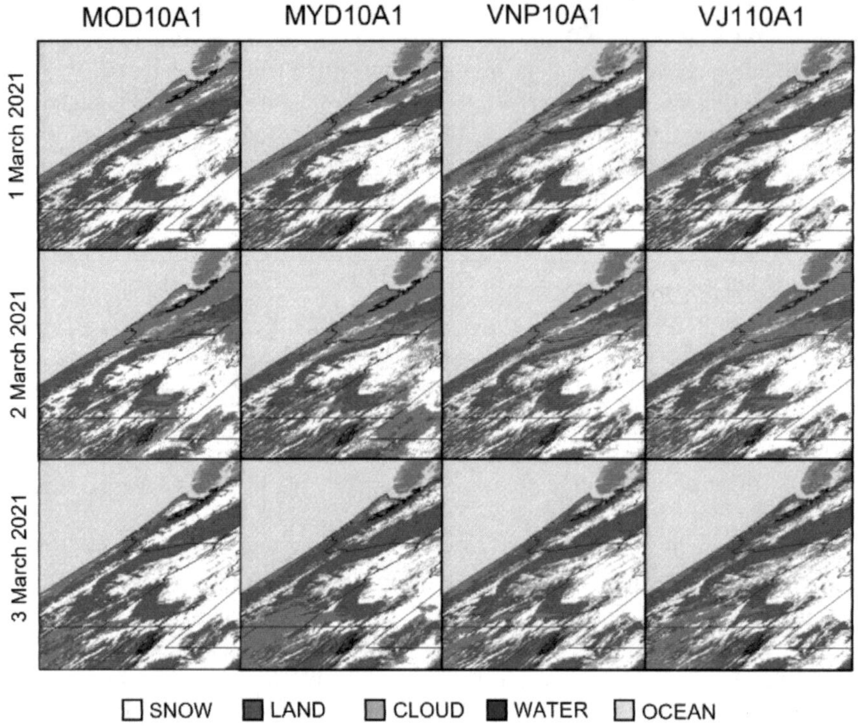

Figure 2. Daily MODIS and VIIRS SCE maps for tile h09v04, 1–3 March 2021. Binary SCE maps on sinusoidal projection.

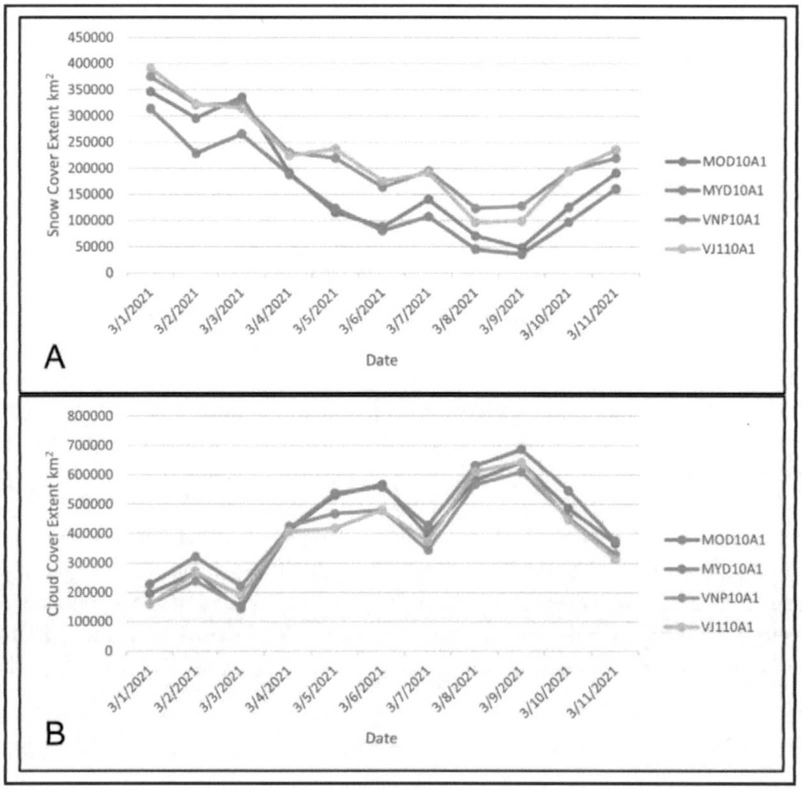

Figure 3. SCE and cloud cover extent of MODIS and VIIRS daily SCE products (*10A1) 1–11 March 2021. SCE is plotted in A, and cloud cover in B.

Comparing SCE in the daily product on a day and tracking SCE day-to-day is confounded by cloud cover. Day-to-day changes in SCE in the *10A1 products are confounded by changes in cloud cover. In this short time series (Figure 3) daily differences in SCE and day-to-day changes in SCE are related to daily and day-to-day changes in cloud over. A day-to-day decrease in SCE (Figure 3A) is associated with an increase in cloud cover e.g., 3–6 March SCE decreases in all SCE maps and cloud cover increases (Figure 3B). Then 9–11 March when SCE increases in all maps and cloud cover decreases (Figure 3A and 3B). The confounding cloud cover needs to be removed so that changes in SCE can be monitored. Similar recommendation to use the CGF products was made in Riggs et al. (2020).

The CGF products (*10A1F) provide a daily "cloud free" SCE to compare SCE maps and to monitor day-to-day changes in SCE. The MODIS and VIIRS CGF SCE products for 1–3 March 2021 are shown in Figure 4. The SCE is very similar spatially and temporally in the MODIS and VIIRS GCF SCE maps and unlike in the daily SCE maps (Figure 2) there is no cloud obscured snow cover. In the absence of cloud cover, SCE is very similar in all the MODIS and VIIRS SCE products every day (Figure 4). Daily comparisons of products and day-to-day changes in CGF SCE are not confounded by cloud cover (Figure 4). Daily SCE in the products on 1–11 March 2021 are plotted in Figure. 5. Over the period, 1–11 March 2021 the trend in SCE shows a slight decrease in all the MODIS and VIIRS maps (Figure 5) which is not apparent in the day-to-day changes in SCE in the daily products (Figure 3A). The SCE areas in CGF SCE (Figure 5) are different from areas for the daily SCE (Figure 3A) because the CGF SCE record began on 1 October 2020 which is the start of the water year and is continuous over the snow season whereas the daily SCE record in this comparison started and 1 March 2021 with cloud cover blocking snow cover that was present on previous days. Clearly, the CGF product is the daily product to utilize for evaluating continuity of SCE among the MODIS and VIIRS products, and for monitoring SCE over a season.

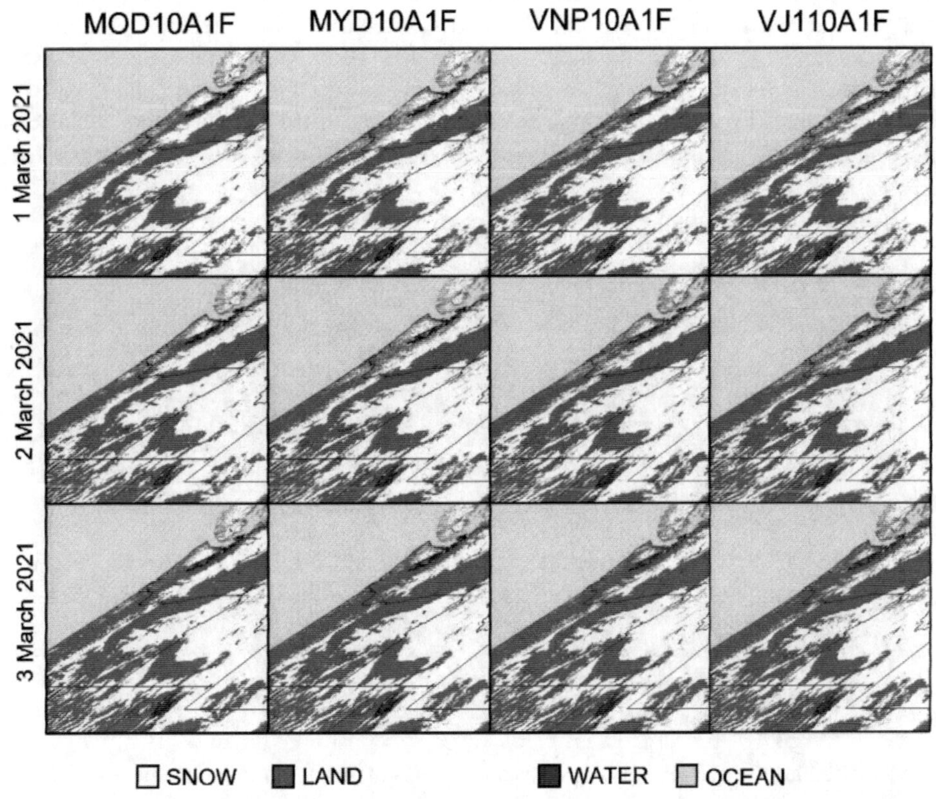

Figure 4. Daily MODIS and VIIRS GCF SCE maps, tile h09v04, 1–3 March 2021.

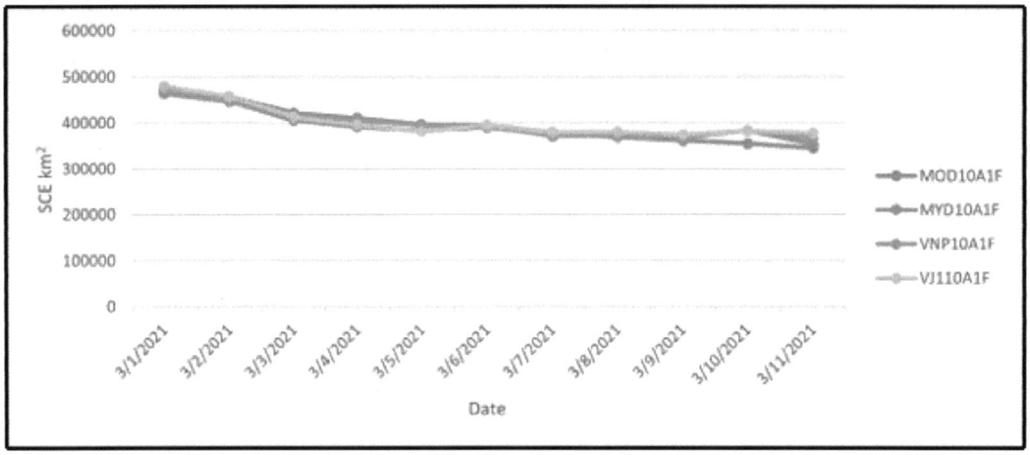

Figure 5. SCE of MODIS and VIIRS daily CGF SCE products (*10A1F) 1–11 March 2021.

Sierra Nevada Region

We investigated continuity of the Terra, Aqua, and SNPP CGF SCE products on the Sierra Nevada Mountain range region (Figure 6) for a winter period 1 December 2022 to 31 January 2023. During that winter period snow cover was continuously present and there were periods of consecutive days with clear views of the surface interspersed with days of cloud cover in the region. SCE extent is very similar in the MODIS and VIIRS daily CGF SCE products throughout this period (Figure 7). In the beginning of the time series, there was an overnight snow event on 1–2 December that deposited snow on the mountains and valleys. It is likely that the more conservative Aqua cloud mask erroneously mapped much of that snow as cloud thus snow was not mapped by Aqua and subsequent days of cloud cover over the region interfered with a return to close agreement in SCE among the three products. The Terra MODIS and SNPP VIIRS data products did not have that same problem with the cloud masking and thus correctly mapped snow from 1–4 December (Figure 7). We think that the cloud-gap filled SCE products are adequately representing the changing snow conditions in this Sierra Nevada region study area (Figure 7).

Figure 6. The Sierra Nevada study region is a 244,530 km^2 area that includes the Sierra Nevada Mountain range, and part of the intermountain region in the western United States. Geographic projection with 500 m grid cell resolution.

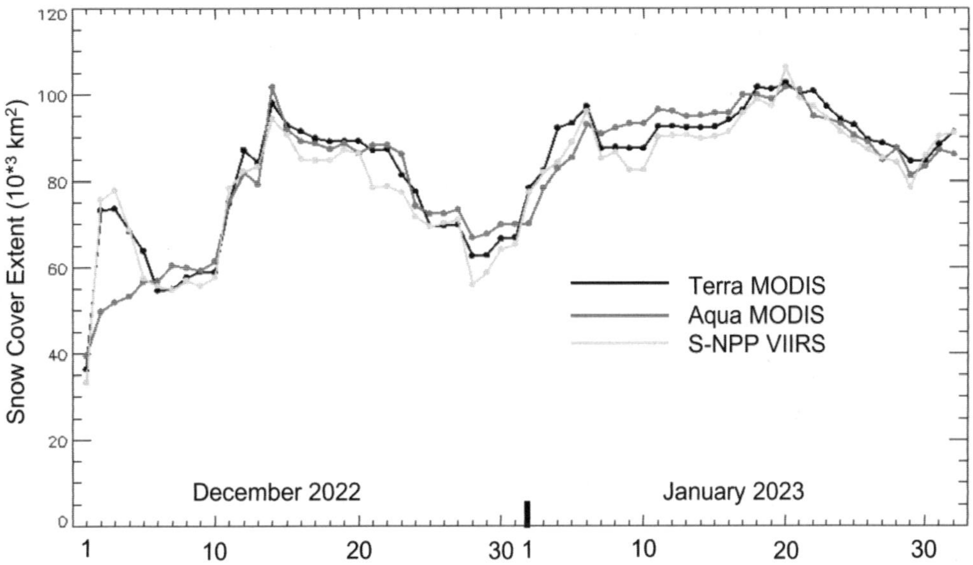

Figure 7. Comparison of daily SCE area among the Terra, Aqua, and SNPP CGF products from 1 December 2022–31 January 2023. Individual plots show results from the Terra MODIS cloud-gap filled (CGF) SCE product, MOD10A1F, the Aqua MODIS CGF SCE product, MYD10A1F, and the S-NPP VIIRS CGF SCE product, VNP10A1F.

Western New York, USA Region

We investigated continuity of the Terra, Aqua, and SNPP CGF SCE products in a region of western New York state region where lake effect snows occur frequently (Figure 8) for a winter period of 1 December 2022 to 31 January 2023. Lake effect snow events occur frequently in this region and cloud cover conditions can change between the morning and afternoon satellite overpasses as well as from day-to-day. Between 1 December 2022 and 31 January 2023, there were very few clear views of the surface. Additionally, a few lake effect snowstorms and ablation events occurred during this period. Timely observation of lake effect snow events, and ablation of snow cover was often not possible because of cloud cover thus, daily SCE is inconsistent among the Terra, Aqua and SNPP products (Figure 9) due to those factors. We do not expect the CGF SCE products can adequately capture the rapidly changing snow conditions in this area of western New York state at this time of year.

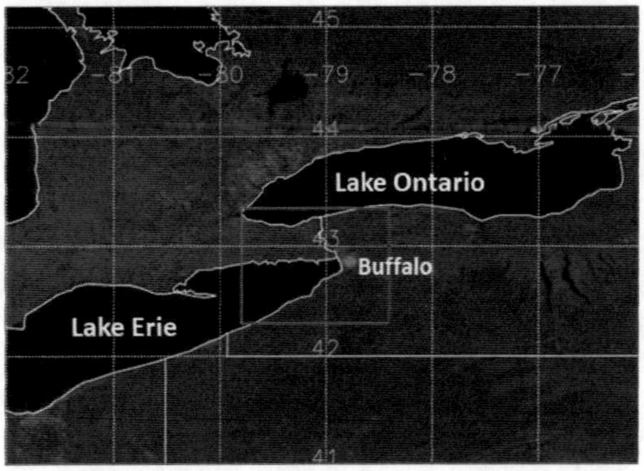

Figure 8. The Buffalo study region is a 13,306 km^2 study area in western New York, US. Geographic projection with 50 m grid cell resolution.

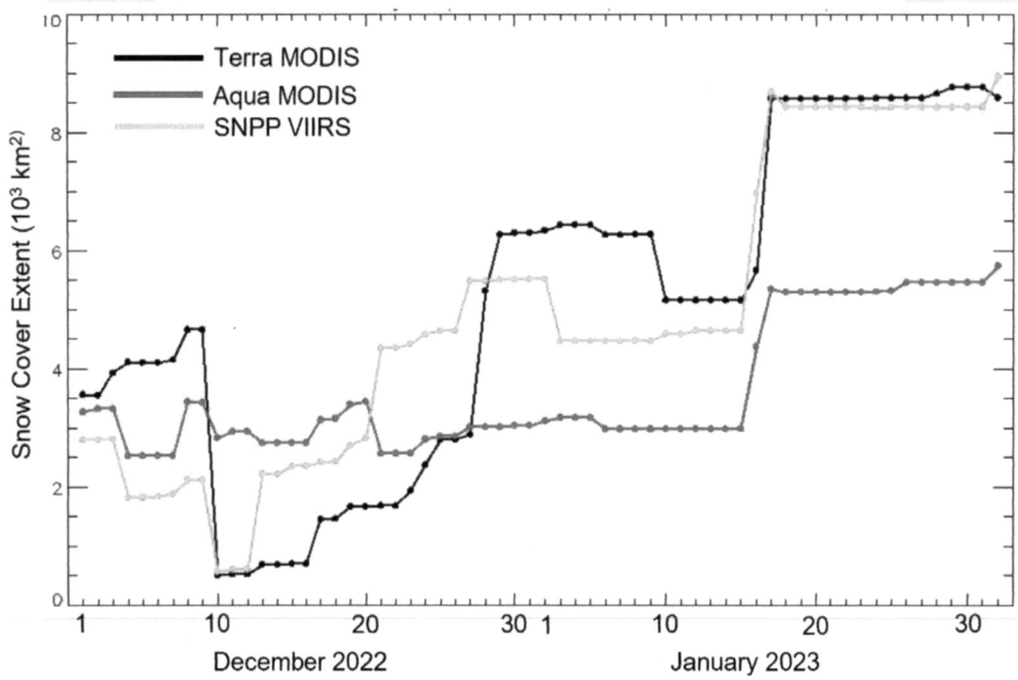

Figure 9. Comparison of SCE among the CGF Terra MODIS MOD10A1F, Aqua MODIS MYD10A1F, and SNPP VIIRS VNP10A1F from 1 December 2022–31 January 2023.

CONCLUSIONS

The daily CGF SCE products are preferable to utilize in investigations of SCE because they minimize the effect of cloud cover, especially for day-to-day monitoring of changes in SCE. SCE extent is very similar in both NASA MODIS and VIIRS SCE products, and trends in SCE are similar as demonstrated in the Pacific Northwest and Sierra Nevada comparisons. A similar result of excellent correspondence in SCE between MODIS and VIIRS SCE products was reported by Hall et al., (2019). VIIRS SCE products are an effective replacement for the MODIS SCE products. There is good continuity between the MODIS and VIIRS daily SCE products. However, there is a caveat, in regions with discontinuous snow cover over a season and frequent cloudy days, e.g., the western New York region the MODIS and VIIRS SCE products may be dissimilar in SCE on any day and large differences in SCE may be observed day-to-day. The MODIS and VIIRS SCE products can be used to build a long-term data record of SCE.

REFERENCES

Hall DK, Riggs GA, DiGirolamo NE, Román MO. 2019. Evaluation of MODIS and VIIRS cloud-gap-filled snow-cover products for production of an Earth science data record. *Hydrology and Earth System Sciences*, **23**: 5227–5241.

Riggs GA, Hall DK, Román MO. 2019a. MODIS Snow Products Collection 6.1 User Guide, Version 1.0, *https://nsidc.org/sites/default/files/c61_modis_snow_user_guide.pdf*. Accessed 21 July 2023.

Riggs GA, Hall DK, Román MO. 2019b. VIIRS/NPP Snow Cover Daily L3 Global 375m SIN Grid, Version C1. Boulder, Colorado USA. NASA National Snow and Ice Data Center Distributed Active Archive Center, *https://doi.org/10.5067/VIIRS/VNP10A1.001*. Accessed 21 July 2023.

Riggs GA, Hall DK. 2021. VIIRS/NPP CGF Snow Cover Daily L3 Global 375m SIN Grid, Version 2. C2. Boulder, Colorado USA. NASA National Snow and Ice Data Center Distributed Active Archive Center, *https://doi.org/10.5067/PN50Y51IVNLE*. Accessed 21 July 2023.

Riggs GA, Hall DK. 2020. Continuity of MODIS and VIIRS Snow Cover Extent Data Products for Development of an Earth Science Data Record. *Remote Sensing*, **12**, 3781. doi:10.3390/rs12223781.

Román MO, Paynter I, et al. 2023. Continuity Between MODIS Collection 6.1 and VIIRS Collection 2 Products. *In preparation*.

Interpreting Cosmic Ray Neutron-Based Snow Water Equivalent Estimates from Heterogenous Snow Distributions

HAEJO KIM[1], MADISON WOODLEY[1], ERIC SPROLES[2], JED EBERLY[3], AND SAM TUTTLE[1]

ABSTRACT

Conventional methods of snow water equivalent (SWE) methods are questionable in the spatially heterogenous snow of prairie environments. Despite low annual snowfall, snowpacks on North American prairies are extensively redistributed by wind, leading to scouring of open areas, and accumulation behind barriers (e.g., trees and fences) or topographically low areas. Cosmic ray neutron sensing (CRNS) techniques are capable of making SWE estimates that are more representative of the areal average than conventional methods due to their large areal footprint, approximately 200-300 meters radius. A CRNS instrument was deployed at the Central Agricultural Research Center in central Montana, USA as a part of the NASA SnowEx field campaign during winter 2020-2021. CRNS has shown to be sensitive to areal average SWE within its footprint in a prairie environment. However, the moderated neutron flux from varying spatial distributions of shallow snowpack within the sensor's footprint remains a source of uncertainty in SWE estimates. To address this uncertainty, we use a Monte Carlo neutron transport model to simulate the CRNS response under varying spatial distribution patterns of snowpack. Preliminary results showed that bare field conditions near the CRNS can yield larger neutron counts, and thus depressed SWE estimates. Once the effects are known, we hope to be able to correct SWE measurements for changes in the snow distribution. These results aid in the interpretation of CRNS SWE estimates in the prairie and demonstrate the value of CRNS observations in highly variable shallow snow environments.

[1] Department of Earth and Environmental Sciences, Syracuse University, Syracuse, NY, USA

[2] Department of Earth Sciences, Montana State University, Bozeman, MT, USA

[3] Agriculture Research, Montana State University, Bozeman, MT, USA

Corresponding author: hkim139@syr.edu

Ku- and L-band SAR Observations of Terrestrial Seasonal Snow and Lake Ice in Ontario during Winter 2023 using the CryoSAR Airborne System

RICHARD KELLY[1], AARON THOMPSON[1], PETER TOOSE[2], CHRIS DERKSEN[2], ALEX ROY[3], LAURA BROWN[4], ALEX LANGLOIS[5], AARON BERG[6], ARVIDS SILIS[2], WEI WANG[1], ZEINAB AKHAVAN[1], JEFF WELCH[1], LINA ZSCHENDERLEIN[1], AND ALEX GELINAS[3]

ABSTRACT

The Cryosphere-Observing SAR (CryoSAR) system is a Ku- and L-band synthetic aperture radar (SAR) system designed to conduct observations of snow and ice on land and water bodies, and soil moisture status. The CryoSAR system is a fully polarimetric SAR with the capability to conduct single or repeat pass observations for interferometric SAR (InSAR) applications. There is significant interest in the Ku (13.5 GHz) polarimetric response and L-band (1.3 GHz) InSAR measurements of snow on land and water to estimate snow water equivalent, a key variable in water resource management applications and in climate change studies. The CryoSAR radars can be operated independently or together. They can also be deployed on a relatively small aircraft, such as a Cessna 208, which is widely available across North America, Europe, and beyond, making the system relatively agile in its deployment. An adjustable mounting system has been designed to enable the instrument to be installed from inside and aircraft and at specified look angles. In fall 2022 and winter 2023, a season-long deployment of the system was conducted in Ontario as part of a Canadian Space Agency-funded project and in support of the Terrestrial Snow Mass Mission. Flights were conducted over selected sites in Ontario including the Haliburton Highlands and Powassan. Field campaigns were also conducted on the ground to provide correlative ground reference data. A combination of traditional field observations of snow properties, and detailed state-of-the-art measurements of microstructure properties were made to quantify the snowpack bulk and stratigraphic characteristics of the snow at the different field sites. This paper presents the initial observations made with the CryoSAR system at Ku- and L-band and correlative field measurements. Results focus on the polarimetric responses from snow on land and on lake ice and demonstrate its applicability for terrestrial snow monitoring.

[1] University of Waterloo, Waterloo, ON, Canada
[2] Environment and Climate Change Canada, Toronto, ON, Canada
[3] Université de Québec Trois-Rivières, Trois-Rivières, QC, Canada
[4] University of Toronto, Toronto, ON, Canada
[5] Université de Sherbrooke, Sherbrooke, QC, Canada
[6] University of Guelph, Guelph, ON, Canada
Corresponding author: rejkelly@uwaterloo.ca

SWE Impact Index: Toward Identifying Critical Regions with SWE Observational Needs

EUNSANG CHO[1,2], CARRIE M. VUYOVICH[1], SUJAY V. KUMAR[1], ETHAN D. GUTMANN[3], BATUHAN OSMANOGLU[1], JAIME BARDAJI[4], PAUL GROGAN[4], AND KWO-SEN KUO[5]

ABSTRACT

Seasonal snow is a vital part of the water cycle, supplying life-giving fresh water to billions of people across the United States and the world. At the same time, large snowpacks with sudden melt events including rain-on-snow events leads to catastrophic extreme flood events, causing large societal and economic consequences. Even though we recognize that human societies in snow-dominant regions are somehow affected by snowpack, an open question still exists: In which regions are more accurate and timely snowpack information critically needed when compared to other regions? Here, we propose a new "SWE Impact index" map showing regions where SWE information is relatively important across the continental U.S. The SWE Impact index is a societal index based on the historical amounts of maximum SWE, rapid snow accumulation and snowmelt, and population. Including population in the index provides a measure of the socio-economic susceptibility for the given areas. In this presentation, we will provide an overview of how we develop the SWE Impact index using a gridded SWE dataset and preliminary results of the index map. This work will help us identify "critical regions" where reliable SWE observations are particularly needed in terms of societal impacts.

[1] Hydrological Sciences Laboratory, NASA Goddard Space Flight Center, Greenbelt, MD, USA
[2] Earth System Science Interdisciplinary Center, University of Maryland, College Park, MD, USA
[3] National Center for Atmospheric Research, Boulder, CO, USA
[4] Stevens Institute of Technology, Hoboken, NJ, USA
[5] Bayesics, LLC, Bowie, MD, USA
Corresponding author: escho@umd.edu

Bias Correction of an Ensemble Mean Reanalysis-Based Permafrost Soil Temperature Product using Snow Cover and Vegetation

TYLER C. HERRINGTON[1], ANDRE R. ERLER[2], AND CHRISTOPHER G. FLETCHER[1]:

ABSTRACT

Reanalysis products provide spatially homogeneous coverage for a variety of climate variables in regions where observational data are limited. However, soil temperature estimates in many reanalysis products are biased cold by 2-7 K across the Arctic, particularly in the cold season. In previous work we showed that a soil temperature product based on the ensemble mean of all available reanalysis products generally outperformed the individual soil temperature estimates of any individual product. Biases over permafrost regions, however, remained relatively large. Here we utilize mean bias subtraction, multiple linear regression, and a random forest regression technique to bias-correct an ensemble mean permafrost soil temperature product. In addition to the thermodynamic state of the land surface in the reanalysis, the random forest regressor also uses information relating to remotely sensed snow cover and vegetation. As part of this process, we develop a vegetation classification scheme that reflects the functional impacts of vegetation on snow cover and soil temperature in order to compare the vegetation cover assumed in the reanalysis products to European Space Agency Climate Change Initiative (ESA-CCI) vegetation cover. We train our models primarily on soil temperature data from Eurasia, as data from North America is sparse. The performance of the correction algorithm will be evaluated in different regions, with specific attention to data-poor regions in North America. The importance of different predictors, and in particular the role of snow biases and vegetation cover will be discussed.

[1] Geography and Environmental Management, Faculty of Environment, University of Waterloo, Waterloo, ON, Canada

[2] Aquanty Inc., Waterloo, ON, Canada

Corresponding author: therring@uwaterloo.ca

Refining and Automating DAV Snow Melt Algorithms using Passive Microwave Calibrated Enhanced-Resolution Brightness Temperature (CETB) Data in Alaska Watersheds

MAHBOUBEH BOUESHAGH[1], JOAN M. RAMAGE[1], VICKI JAGDEO[1], AND MARY J. BRODZIK[2]

ABSTRACT

Snowmelt is vital in snow-covered regions as it provides water for groundwater recharge, agriculture, industry, and domestic use. Understanding the timing and magnitude of snowmelt is essential for the water cycle, seasonal distribution of runoff, and disaster management (wildfire and flooding). For years, passive microwave (PMW) brightness temperatures (Tb) have been used to identify melt-freeze dynamics in snowpack. Liquid water in snow causes an increase in emissivity, drastically increasing Tb, which can be detected through PMW remote sensing. The research employs the newly available Calibrated, Enhanced-Resolution Brightness Temperature (CETB) dataset to estimate the timing of snowmelt at higher spatial resolution (~3-6 km). Melt detection algorithms will be refined using the diurnal amplitude variation (DAV) approach established by Ramage *et al.* We will determine and automate the range and optimum Tb and DAV thresholds (legacy thresholds of 246K / ±10K for Special Sensor Microwave Imager or SSM/I) in Alaska watersheds using existing and supplemental observational and reanalysis datasets as well as different mathematical analysis, such as histogram and time series analysis, and derivatives. We expect improved melt timing and a range of thresholds sensitive to different terrain and environments. We will compare algorithm results with ground-based measurements compiled from multiple sources, such as air and snow temperatures, snow water equivalent (SWE), snow depth, and stream discharge. This study offers an understanding of how the use of higher-resolution CETB data and automatic approaches can enhance snowmelt analysis, granting hydrologists the ability to examine timing and patterns of snowmelt across different regions and snow types.

[1] Department of Earth & Environmental Sciences, Lehigh University, Bethlehem PA, USA

[2] NSIDC/CIRES, University of Colorado Boulder, Boulder, CO, USA

Corresponding author: mabb21@lehigh.edu

Comparing Passive Microwave Snowmelt Detection Methods using Ground-Based Snowmelt Observations

ANGELA RIENZO[1], SAMUEL TUTTLE[1], GEORGE DUFFY[1], ELIAS DEEB[2], AND CARRIE VUYOVICH[3]

ABSTRACT

Melting snow is an integral factor in understanding both the hydrologic cycle and energy budget of many regions. Monitoring wet or melting snowpacks is vital for river flood predictions, water resource management, monitoring climatic trends and changes, and avalanche risk forecasting. It is also important for researchers and government agencies monitoring snow depth and snow cover using passive microwave satellites, as liquid water in snow can cause high brightness temperatures due to differences in dielectric properties between wet and dry snow. Sites that collect continuous ground measurements of snow liquid water content are rare, while satellite observations are easily accessible and have the potential for snowmelt monitoring on a global scale. The goal of this study is to compare and evaluate the information provided by different snowmelt detection methods that use passive microwave satellite observations. Certain algorithms may be more sensitive to certain types of melt events, such as midwinter melt/refreeze events or spring melt onset/duration. For example, previous work has shown that the DAV (diurnal amplitude variation) method is best at showing mid-winter and melt onset events, while it is less useful for detecting late-stage spring melting. This study extends that research by applying similar analysis methods to different snowmelt detection algorithms. Snowmelt events detected using various algorithms applied to AMSR2 passive microwave brightness temperature data are compared to ground-based snowmelt observations, including continuous liquid water content measured by a Snow Pack Analyzer-2 located within the Sleeper's River Research Watershed in Vermont.

[1] Department of Earth and Environmental Sciences, Syracuse University, Syracuse, NY, USA
[2] U.S. Army Cold Regions Research and Engineering Laboratory, Hanover, NH, USA
[3] Hydrological Sciences Laboratory, NASA Goddard Space Flight Center, Greenbelt, MD, USA
Corresponding author: arrienzo@syr.edu

Multidisciplinary Observatory for Arctic Climate Change and Extreme Events Monitoring (MOACC)

DANIEL KRAMER[1,2]

ABSTRACT

The Multidisciplinary Observatory for Arctic Climate Change and Extreme Events Monitoring (MOACC) proposal is submitted by the Université de Sherbrooke (lead institution-UdeS), University of Toronto (UofT), Western University (WU) and Université de Montréal (UM). The main objective of our project is to develop a permanent multidisciplinary scientific infrastructure that enables long-term observations of Arctic climate change, bringing together experts from a wide range of expertise and institutions. The project is led by Prof. Alexandre Langlois (UdeS) and Prof. Kimberley Strong (UofT) and responds to a consensus on the lack of temporal observations that are crucial to understand feedback processes and to promote model development in the Arctic. The innovative aspect of this proposal resides in its multidisciplinary approach while enabling long-term Arctic measurements spanning several disciplines. The proposed observatory will be located at the Canadian High Arctic Research Station (CHARS) in Cambridge Bay, Nunavut, while enhancing the reach of CHARS with linkages to the Environment and Climate Change Canada supersite in Iqaluit. Our ambition is to establish the site as one of the largest instrumented high Arctic observatories dedicated to the monitoring of key indicators that drive climate change. The site will generate and enhance partnerships, not only with Canadian research centers and organizations, but also with international research partners and networks.

[1] Groupe de Recherche Interdisciplinaire sur les Milieux Polaires (GRIMP), Université de Sherbrooke, Sherbrooke, QC, Canada

[2] Centre d'études nordiques (CEN), Université Laval, Québec City, QC, Canada

Corresponding author: daniel.kramer@usherbrooke.ca

Oral Session #4

Session Chairs: Sam Tuttle & Alicia Pouw

How Representative are Low Resolution Sea Ice Concentration Products of Conditions at Coastal Sites along the Central Western Antarctic Peninsula?

ANDREW G. KLEIN[1], CARLY K. CASPER[2], JESSICA N. FITZSIMMONS[1], DARREN H. HENRICHS[1], CHARLES D. AMSLER[3], MARGARET O. AMSLER[3], KATRIN IKEN[4], AARON W.E. GALLOWAY[5], SABRINA HEISER[6], ALEX LOWE[7], JULIE .B. SCHRAM[8], AND ROSS WHIPPO[5]

ABSTRACT

Macroalgal forests exist along the northern portion of the western Antarctic Peninsula (WAP). However, few studies have documented its distribution farther south in the central WAP. In 2019, macroalgal cover was determined at 14 coastal sites located along a latitudinal gradient (64°-69° S) through quantitative analysis of diver-collected video from replicated vertical transects collected at depths between 40 m and 5 m. A strong negative correlation was documented between total macroalgal cover and all the sea ice concentrations evaluated (National Ice Center Charts, AMSR-E and AMSR-2 and the National Snow and Ice Data Center Sea Ice Index). However, these remote-sensing derived sea ice concentrations are captured at large spatial scales relative to the scale of the sampling sites (~100 meters across). To investigate how representative these sea concentrations are of conditions at the studied shallow water coastal sites, sea ice concentrations from Landsat (30 m) and MODIS (1000 m) were computed for buffers ranging from 100 m 10,000 m around each site and then compared to concentrations from the larger footprint products. In general, the larger footprint sensors indicated lower sea ice concentrations than Landsat or MODIS. However, the observed differences at individual sites were typically consistent across buffer sizes. This indicates that while the sites had higher sea ice concentrations than retrieved from the lower resolution products, the higher concentrations appear to be consistent across spatial scales from 100s to 1000s of meters.

[1] Texas A&M University, College Station, TX, USA
[2] Central College, Naperville, IL, USA
[3] University of Alabama at Birmingham, Birmingham, AL, USA
[4] University of Alaska Fairbanks, Fairbanks, AK, USA
[5] Oregon Institute of Marine Biology, University of Oregon, Charleston, OR, USA
[6] Marine Science Institute, University of Texas-Austin, Port Aransas, TX, USA
[7] Tennenbaum Marine Observatories Network, Smithsonian Institute, Edgewater, MD, USA
[8] University of Alaska Southeast, Juneau, AK, USA
Corresponding author: klein@tamu.edu

Validation of Snow Water Equivalent Products: Dialed in for Non-Mountain Regions but Challenges Remain in Complex Terrain

COLLEEN MORTIMER[1], LAWRENCE MUDRYK[1], EUNSANG CHO[2,3], CHRIS DERKSEN[1], AND CARRIE VUYOVICH[2]

ABSTRACT

We advance the evaluation of gridded snow water equivalent (SWE) products by using reference data from both snow courses and airborne gamma-derived SWE estimates to evaluate fourteen datasets over North America. These products cover a broad range of snow analysis from those based on Earth-observation data (JAXA-AMSR2, Snow CCI SWE v1 and v2), coupled land-atmosphere reanalysis (ERA5, ERA5-Land, ERA5-Snow, GLDASv2.2, MERRA2, JRA-55), snow models of varying complexity driven by reanalysis data (Crocus driven by ERA-Interim and ERA5, Brown Temperature Index Model (TIM) driven by ERA5 and JRA-55), and data assimilation schemes (U. Arizona). We assess product performance across both non-mountainous and mountainous regions and analyze the sensitivity of this performance to the choice of reference data set. In non-mountain areas, product performance is insensitive to the choice of reference dataset and there is strong agreement of reference SWE up to spatial scales of at least 25 km, comparable to the grid spacing of most coarse resolution products. In mountain areas, there is poor agreement between the reference datasets even at short distances (< 5 km) and the choice of reference dataset strongly influences the sign and magnitude of the product errors (bias and RMSE) due to systematic elevation differences among the reference data locations and the product grid cell centroids. Minimizing these systematic elevation-driven biases is essential to a fair evaluation of the product performance and improves the median RMSE of the model-based products by one third and almost 50% using gamma and snow courses, respectively. Our full analysis demonstrates that we can robustly validate product SWE estimates in non-mountain regions and are able to provide relative rankings of product performance in mountain regions.

[1] Environment and Climate Change Canada, Toronto, ON, Canada
[2] Hydrological Sciences Laboratory, NASA Goddard Space Flight Center, Greenbelt, MD, USA
[3] Earth System Science Interdisciplinary Center, University of Maryland, College Park, MD, USA
Corresponding author: colleen.mortimer@ec.gc.ca

Assimilation of Airborne Gamma-Ray Observations Provides Utility for SWE Estimation in Forested Environments of the Northeastern United States

EUNSANG CHO[1,2] , YONGHWAN KWON[1,2], SUJAY V. KUMAR[1], AND CARRIE M. VUYOVICH[1]

ABSTRACT

An accurate estimation of snow water equivalent (SWE) has been a critical issue for water supply predictions and snowmelt-driven flood management. Data assimilation (DA) is a promising approach to improve modeled SWE estimations at a large spatial scale by merging remote sensing measurements into model predictions. In forested regions, however, most remote sensing techniques for estimating snowpack (e.g., passive microwave, photogrammetry, & lidar) is hampered by the effects of the forest canopy, resulting in large uncertainties in the SWE estimations. A well-established, but little-known airborne gamma radiation technique has provided a strong potential to estimate snowpack conditions in forested environments because the gamma-ray technique uses a difference in gamma-ray particles between snow-off and snow-on conditions by limiting canopy effects. In this study, we assimilated airborne gamma-based SWE retrievals into the Noah land surface model with multi-parameterization options (Noah-MP) in the Northeastern U.S. via the NASA Land Information System framework. Our results show that assimilation of the airborne gamma ray SWE observations improved the modeled SWE despite the limited number of gamma SWE observations (up to only four observations during the winter period) even in densely forested regions.

[1] Hydrological Sciences Laboratory, NASA Goddard Space Flight Center, Greenbelt, MD, USA
[2] Earth System Science Interdisciplinary Center, University of Maryland, College Park, MD, USA
Corresponding author: escho@umd.edu

Improving our Understanding of ICESat-2 Ice Thickness Estimates in the Canadian Arctic Archipelago using *in situ* and Drone Measurements

PETER TOOSE[1], JOSHUA KING[1], MIKE BRADY[1], COLLEEN MORTIMER[1], BENOIT MONTPETIT[1], CHRIS DERKSEN[1], STEPHEN HOWELL[1], AND ARVIDS SILIS[1]

ABSTRACT

The Canadian Arctic Archipelago (CAA) is a unique sea ice region located across northern Canada, adjacent to the Arctic Ocean. To date, monitoring of sea ice within the CAA has lacked volumetric / ice thickness information. NASA's Ice, Cloud, and land Elevation Satellite-2 (ICESat-2) was launched in 2018 and included a key objective to estimate sea ice thickness. The Advanced Topographic Laser Altimeter System (ATLAS) aboard ICESat-2 measures surface elevation at high resolution with a footprint diameter of approximately 11 m. ATLAS was designed to obtain accurate and routine estimates of sea ice surface height and freeboard (ice plus snow thickness) relative to the height of the adjacent ocean. However, open water within the CAA is scarce due to high concentrations of *in situ* formed landfast ice with few open water leads, and snow thickness is generally unknown. To better understand these uncertainties, researchers from Environment and Climate Change Canada conducted surveys on the sea ice between April 10th to 28th, 2022, documenting sea ice surface heights, snow depth, freeboard and ice thickness using a combination of manual, global navigation satellite system (GNSS) surveys, and drone measurements. Surveys were conducted along 5 ICESat-2 satellite overpasses, at 57 sites, and along 86 km of continuous transects with coincident snow depth and EM-31 electromagnetic induction-sensed sea ice thickness observations. Comparisons of our *in situ* measurements with the ICESat-2 satellite products for sea ice elevation (ATL07), and sea ice freeboard (ATL10) will be presented.

[1] Environment and Climate Change Canada, Toronto/Ottawa, ON, Canada
Corresponding author: peter.toose@ec.gc.ca

Ruminations on Machine Learning and Snow Mass

BARTON A. FORMAN[1], GOUTAM KONAPALA[2], AND SUJAY V. KUMAR[3]

ABSTRACT

Machine learning applications, with particular relevance to terrestrial snow, have received a lot of attention in recent years. Despite all the efforts to harness machine learning in the prediction of snow depth (or SWE), many challenges remain. This presentation highlights commonly used machine learning techniques and will discuss the many perils and pitfalls.

A suite of commonly-used algorithms – support vector machine regression (SVM), Gaussian process regression (GP), ensemble learners (EL), long short-term memory networks (LSTM), and gated recurrent networks (GRU) – focused on time series regression will be presented. The machine learning predictions presented here rely on a mixture of satellite-based measurements of passive microwave radiometry from AMSR-E and AMSR2 or advanced land surface model output from the NASA Land Information System (LIS) as the inputs. All algorithms discussed here employ supervised learning using either ground-based observations from SNOTEL (geophysical space) or passive microwave brightness temperature (observation space) serving as the training targets.

Examples of what does work, and more importantly, what does *not* work, will be presented. This presentation will show that machine learning is not, despite all the marketing, magic. Machine learning is merely applied math. Furthermore, it will be argued that machine learning is not a panacea. However, as more and more snow researchers find machine learning increasingly useful in their research, the trend of machine learning adoption will continue to increase as "big data" grows ever bigger. The goal of this research is to help share ideas and enhance community understanding of machine learning algorithms as applied to terrestrial snow in pursuit of a solution to an ill-posed, underdetermined problem that has vexed the snow community for decades.

[1] Department of Civil and Environmental Engineering, University of Maryland, College Park, MD, USA
[2] Hydrological Sciences Laboratory, NASA Goddard Space Flight Center, Greenbelt, MD, USA
Corresponding author: baforman@umd.edu

Snow Water Equivalence and Stratigraphy Records from White Glacier, Axel Heiberg Island, Nunavut: 1959-2023

LAURA THOMSON[1] AND MILES ECCLESTONE[2]

ABSTRACT

White Glacier on Axel Heiberg Island, Nunavut, is one of seven Canadian reference glaciers to the World Glacier Monitoring Service, with near-continuous mass balance observations spanning 1959-2023. While an important indicator of climate-driven changes in this region of the Arctic, these data also comprise a valuable record of snow accumulation and distribution patterns thus presenting an opportunity to investigate temporal trends and spatial patterns in regional snowfall over the last 60+ years.

The winter snowpack on White Glacier is measured annually in April/May at approximately 15 locations and includes measurements of snow water equivalence (SWE) and stratigraphy (Figure 1). From 1959 to 2022, we calculate an average annual winter snowpack SWE of 205 ± 114 mm w.e. from snow pits above 800 m a.s.l. (n=748). While accumulation increases weakly with elevation (approx. 54.5 mm w.e./km), SWE observations at the highest sites are found to be notably lower than the average accumulation rate of 371 mm w.e. a^{-1} measured in a snow/firn shaft spanning the years 1929-1962 on nearby Müller Ice Cap. No statistically significant trend in snowfall has been detected from these *winter* (September-May) snowpacks to support a hypothesized "warmer, wetter Arctic" at this location. However, mass balance stake measurements and snow depth sounder data suggest *summer* accumulation should no longer be considered negligible, as was once thought, in the calculation of glacier mass balance.

Figure 1. Snow pit analysis at White Glacier in 2012, expertly demonstrated by Miles Ecclestone.

[1] Department of Geography and Planning, Queen's University, Kingston, ON, Canada
[2] Department of Geography (formerly), Trent University, Peterborough, ON, Canada
Corresponding author: l.thomson@queensu.ca

Oral Session #5

Session Chairs: Tyler Herrington & Eli Deeb

Using Lakes as Snow Pillows: Monitoring Snowfall from Lake Water Pressure in the Adirondack Mountains, NY

SAMUEL E. TUTTLE[1], COLIN M. BEIER[2], JAMES S. MILLS[3], AND HAMISH D. PRITCHARD[4]

ABSTRACT

Snowfall is a fundamental winter weather observation that is used for benchmarking weather models, forcing snowpack models, and tracking climate change. However, accurate measurement of snowfall is challenging, due to precipitation gauge undercatch, wind redistribution, and other local effects. A recently published method (Pritchard et al., 2021, J. Hydrometeorology) used water pressure measured in lakes during the winter to estimate the mass of snow falling onto the lake surface. The rationale behind this method is simple: if snow falls on to the lake surface, the water pressure in the lake will increase, either by additional weight on top of the ice or by adding water to lake. This method has the potential to increase the footprint over which snow is measured by multiple orders of magnitude, because the entire lake surface acts as a giant snow pillow, reducing measurement errors that can arise in conventional, point-scale snowfall measurements. However, the lake snowfall method has not been broadly tested, and only at high latitude and high altitude. We evaluate the viability of the method in a location with more variability in winter weather, including midwinter melt events, to determine whether it can be applied in regions with temperate climates. Specifically, we test the method in Arbutus Lake, located within a long-term ecological monitoring site in the Adirondack Mountains of New York, in winter 2022-2023. Lake-estimated snowfall is compared to nearby snow and meteorological observations, and we supplement the method with streamflow measured at the lake inlet and outlet.

[1] Department of Earth and Environmental Sciences, Syracuse University, Syracuse, NY, USA

[2] State University of New York College of Environmental Science and Forestry, Syracuse, NY, USA

[3] State University of New York College of Environmental Science and Forestry Newcomb Campus, Newcomb, NY, USA

[4] British Antarctic Survey, Cambridge, United Kingdom

Corresponding author: setuttle@syr.edu

Comparative Analysis of NOAA and NASA Snow Cover Extent Products

DAVID A. ROBINSON[1], THOMAS W. ESTILOW[1], DOROTHY K. HALL[2], AND GEORGE A. RIGGS[3]

ABSTRACT

Key satellite-derived snow cover extent (SCE) Earth Science Data Records (ESDR) from NOAA and NASA are invaluable in monitoring SCE variability and change. However, questions remain regarding specific SCE values, including confidence limits for these climate ESDRs. Here, we report on a new effort to achieve a better understanding of each product's strengths and weaknesses to ensure the best SCE ESDR at meso-to-macro scale levels is available to the climate community and others in need of such valuable information. Analyses are underway utilizing NOAA/IMS and NASA MODIS and VIIRS products on regional and seasonal levels over the past two decades. In addition to cases studies for individual regions and dates, project analyses include longitudinal evaluations over a wide range of landscapes. Preliminary results for selected regions in all seasons are providing encouraging spatial correlations between products, though fall extents are found to differ between the NOAA and NASA products more than winter and spring extents. This is likely due to a combination of more extensive cloud cover and lower solar illumination during fall. Ultimately, study results will direct users to a particular map product depending upon their needs with regard to spatial resolution or temporal continuity.

INTRODUCTION

While considerable attention has been paid to specific characteristics associated with sea ice, glacial ice, and ice sheets, including variability and trends, less attention has been paid to the distribution of snow cover in the climate system. This is not to say that snow cover has been ignored, however snow on land, the climate variable exhibiting the greatest intra-annual variability, requires additional scrutiny. This includes assessment of the extent of snow cover, where it has long been recognized that spring change is in progress (Robinson and Dewey, 1990). Declining snow cover extent (SCE) has been particularly notable across the Arctic and Sub-Arctic in May and June over the past decade. On the other hand, sizeable increases in SCE have been noted over northern interior lands in mid-late fall. This is apparently associated with enhanced moisture fluxes and persistent circulation regimes, both of which may be influenced by greater open water in the Arctic (Allchin and Déry, 2020).

[1] Rutgers University, Piscataway, NJ, USA
[2] Earth System Science Interdisciplinary Center / University of Maryland, College Park, MD, USA
[3] Science Systems and Applications, Inc., Lanham, MD, USA
Corresponding author: david.robinson@rutgers.edu

Key satellite-derived SCE Earth Science Data Records (ESDRs) from NOAA and NASA have been invaluable in monitoring variability and change (Estilow et al., 2015; Woods and Helfrich, 2019; Hall et al., 2019; Riggs and Hall, 2020). These products continue to be the preeminent SCE ESDRs in existence and are likely to remain so for the foreseeable future. However, questions remain regarding specific SCE estimates, including confidence limits for these climate ESDRs. To address these issues and gain a strong understanding of each product's strengths and weaknesses, our team is conducting comparative analyses. Initial results are presented here. Ultimately, this will lead to the best Northern Hemisphere continental SCE ESDR at meso-to-macro scale levels available to the climate community and others in need of such valuable information.

This current study serves as a precursor to a future one that will combine these products into an ESDR with meaningful uncertainty bounds that will extend from the start of the MODerate-resolution Imaging Spectroradiometer (MODIS) era through the present and future Visible Infrared Imaging Radiometer Suite (VIIRS) era, during which time the NOAA IMS analysis/mapping system has been employed to generate the NOAA daily product. Furthermore, combining the NASA MODIS and VIIRS SCE products with the weekly NOAA product that the Rutgers team recently re-digitized at 24 km resolution from 1981–1999, will yield a more confident CDR that can be extended back two decades further. While NOAA weekly maps are available from late 1966, the quality prior to the 1980s is such that re-digitizing to 24 km was not considered useful.

NASA AND NOAA SNOW COVER EXTENT PRODUCTS

We use the NASA MODIS Terra cloud-gap filled (CGF) normalized-difference snow index (NDSI) daily snow-cover product, MOD10A1F, Collection 6.1. A clear view of the surface is provided each day by looking back to the last cloud-free observation of the surface on a per-pixel basis (Hall et al., 2010 & 2019 and Riggs et al., 2018). The MOD10A1F product was developed and produced at Goddard Space Flight Center (Riggs et al., 2017) and is distributed by the National Snow and Ice Data Center (https://nsidc.org/data/mod10a1f/versions/61) as part of a suite of NASA standard snow-cover products [https://modis-snow-ice.gsfc.nasa.gov/]. Daily MOD10A1F data products are available beginning 24 February 2000 (Fig. 1), and extending to the present, except during short-term outage periods caused by spacecraft maneuvers, temporary software or instrument issues or upgrades. The snow cover products have been validated at Stage 2, meaning that product accuracy is estimated over a significant (typically > 30) set of locations and time periods by comparison with reference in situ or other suitable reference data. The greatest uncertainty in SCE mapping is cloud obscuration of the surface.

The Interactive Multisensor Snow and Ice Mapping System (IMS) is a software tool used for daily monitoring of snow and ice extent across the Northern Hemisphere. Qualified analysts at the U.S. National Ice Center assess snow and ice cover using a combination of satellite imagery, derived products, and surface observations. Prior to the charting process, products and imagery from various sources are pre-processed to match the IMS system's geographic parameters (Fig. 2). These data are then retrieved and displayed on IMS workstations, where analysts manually delineate snow covered lands. IMS mapping begins using the previous day's SCE output, which is modified based on interpretation of the most recent data available in the system. According to Helfrich et al. (2007), looping visible satellite imagery in combination with meteorological data comprise 80 to 90% of the inputs used for SCE analysis. IMS output at 24-km nominal resolution has been produced operationally since November 1998, and is considered valid at midnight UTC each day.

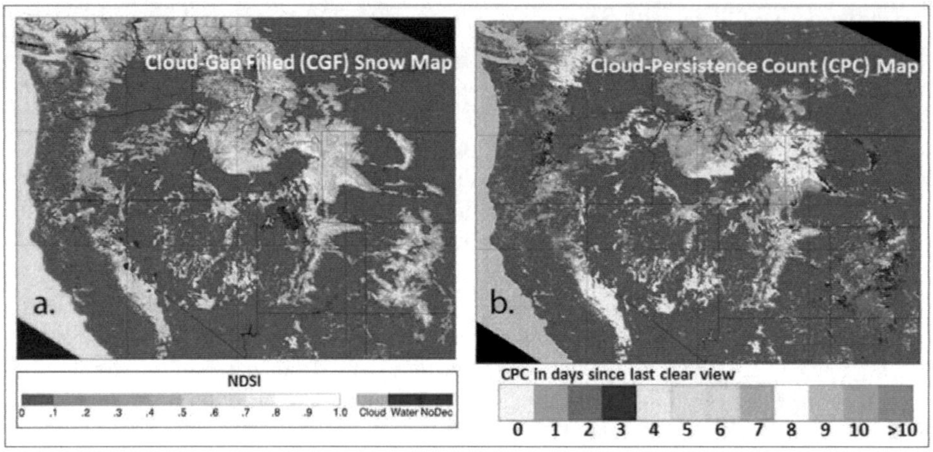

Figure 1. (a) Terra MODIS cloud-gap-filled (CGF) MOD10A1F snow map for 19 March 2012. (b) Cloud-persistence count (CPC) map associated with the MOD10A1F snow map from the quality assurance (QA) data set for the CGF snow map in (a). For 19 March 2012, when a pixel has a CPC of zero, it means that the NDSI value for that pixel was acquired on 19 March 2012. When a pixel has a CPC of one, it means that the NDSI pixel value is 1 day old, and was therefore acquired on 18 March, and so on (from Hall et al., 2019).

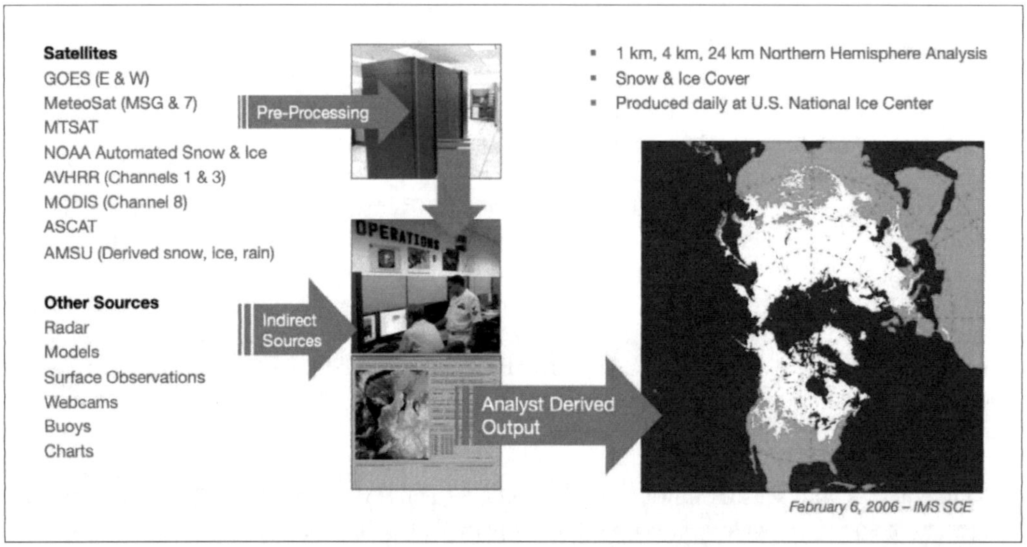

Figure 2. IMS generation process from input sources to processing to daily maps (Helfrich, personal communication.).

STUDY APPROACH

The Rutgers team has assembled a time series of 24-km resolution daily IMS snow cover extent maps for the 2002–2020 study interval. These maps cover the Northern Hemisphere and are quality-controlled for consistency with a standardized land mask applied. Aggregation to monthly maps and SCE area calculations have been performed according to the Rutgers routine (Robinson et al, 1993). The NASA team has developed a time series of MODIS cloud-gap filled snow maps of North America and Eurasia. Comparative studies have been undertaken at continental, regional, and sub-regional scales.

With the focus on product comparisons, particular interest has been paid to times and regions where SCE results differ among the products. This includes employing ancillary data sources such as in situ station observations, Landsat imagery, and the NASA and NOAA imagery that was employed in generating the snow products. This is not an exercise to determine which product is better, as it is known that each has its strengths and weaknesses. Rather, as results illustrate below, it has been possible to identify regions and seasons where differences have occurred.

RESULTS

Initial "blind" comparisons between NOAA IMS and NASA MODIS daily SCE maps find large differences between the two in early and mid-season. Figure 3 exemplifies this for the 2018/2019 and 2019/2020 seasons. Once snow coverage is mapped, it is immediately apparent that the situation results from the MODIS product not mapping snow in high latitude lands deemed to be lacking sufficient solar illumination for the snow detection algorithm to work, thus those regions are shown as dark or missing. Meanwhile, the IMS manual mapping approach can identify snow cover under these conditions or assumes that boreal lands too dark for recognition are snow covered (Fig. 4). As Helfrich et al. (2007) report: "Analysts rely more on snow climatology to estimate snow cover in the high latitudes during the winter than pure microwave data… Even during winter, microwave derived snow data generally only represents 5% of an analysis."

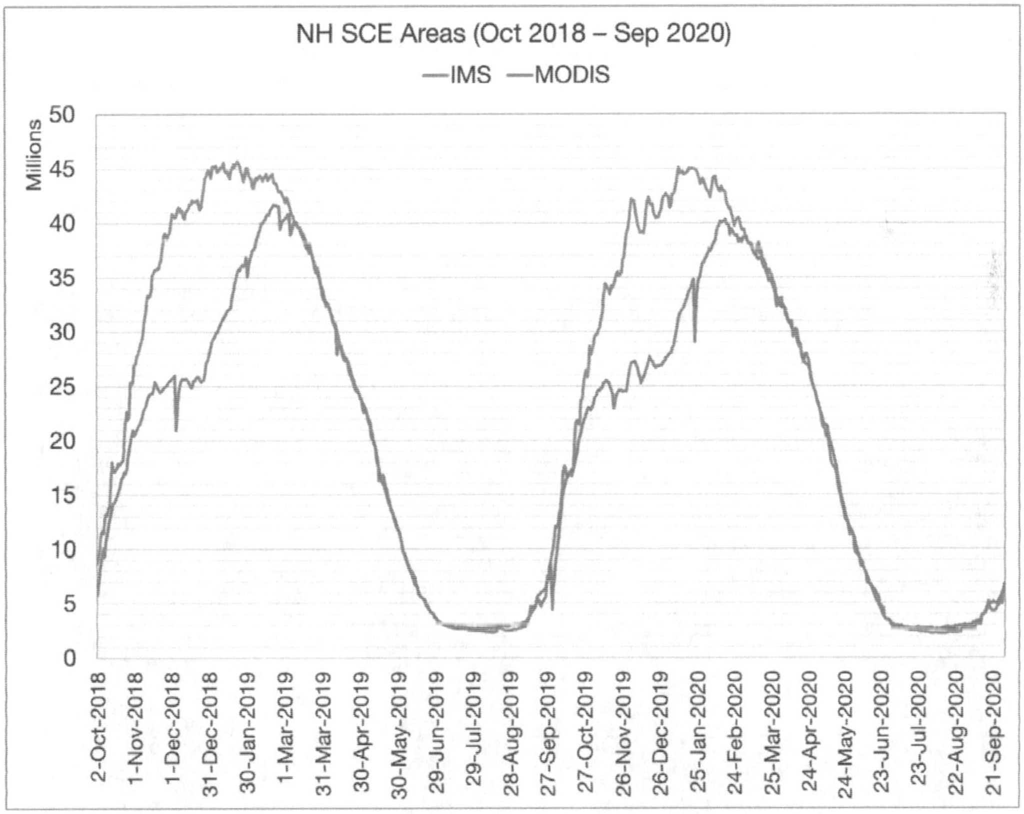

Figure 3. SCE time series for MODIS and IMS products from October 2018 to September 2020.

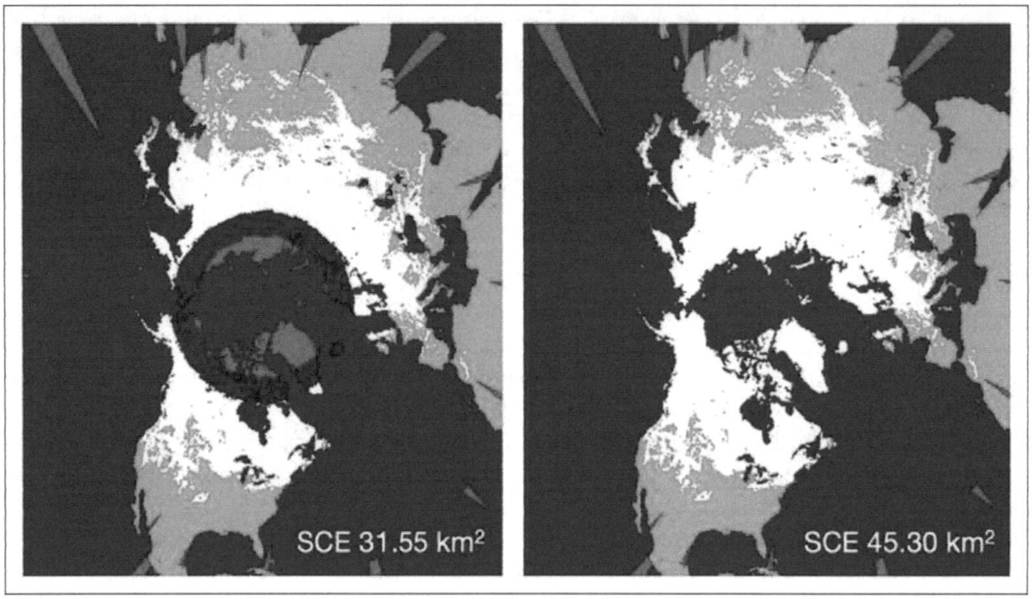

Figure 4. MODIS map of SCE on 10 January 2006 with polar night area in red (left) and with it removed, assuming masked land areas are fully snow covered (right).

Assuming that the dark region on the MODIS map is snow covered results in maps of the two products that are more closely matched (Fig. 5). This is confirmed when comparing the 2018/2019 seasonal SCE mapped by each product (Fig. 6). The close match between MODIS and IMS hemispheric SCE, both in space and over time, provides a strong level of confidence for the overall accuracy of each mapping routine at a macro scale. However, in most years the IMS product shows slightly higher SCE in the early and mid-fall with a closer match in spring. This varies somewhat from year to year, with the 2017/2018 season showing the largest difference between the two over the ten years that were compared (Fig. 7).

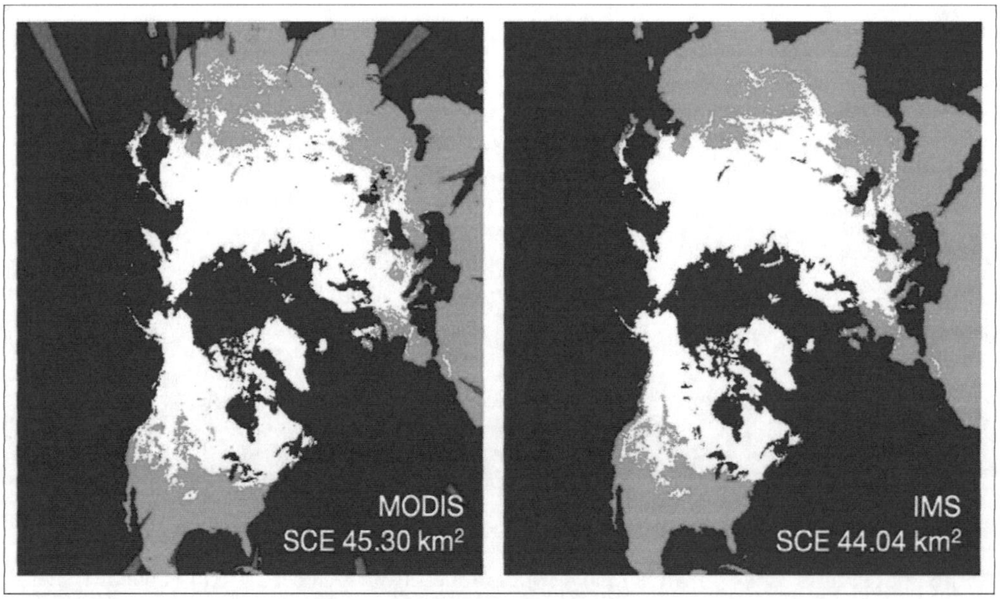

Figure 5. Adjusted MODIS map of SCE on 10 January 2006 from figure 4 (left) and IMS SCE map for that day (right).

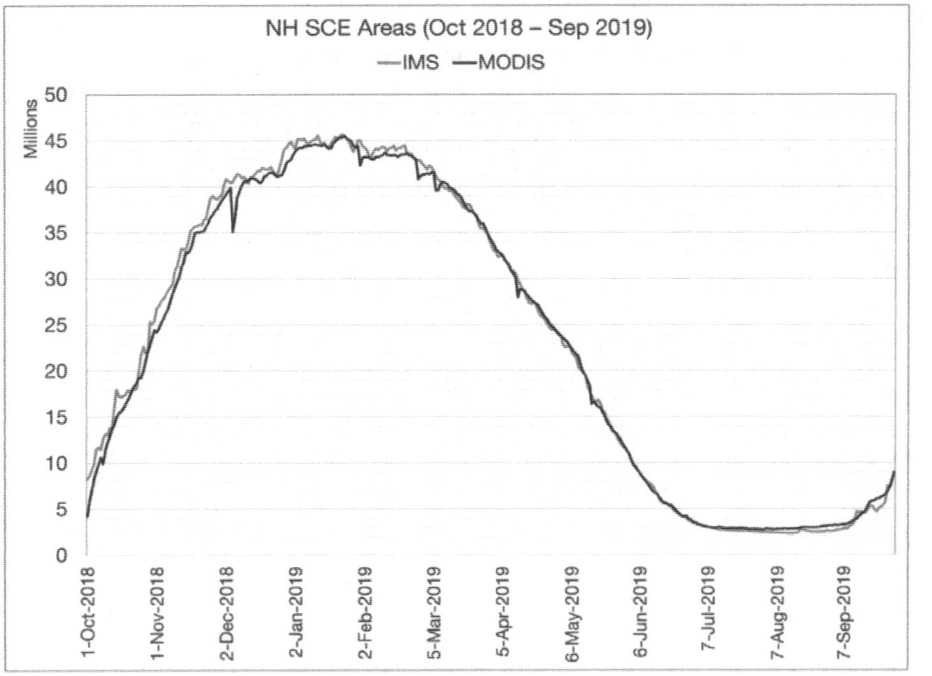

Figure 6. MODIS and IMS Northern Hemisphere SCE from October 2018 to September 2019 after polar night adjustment.

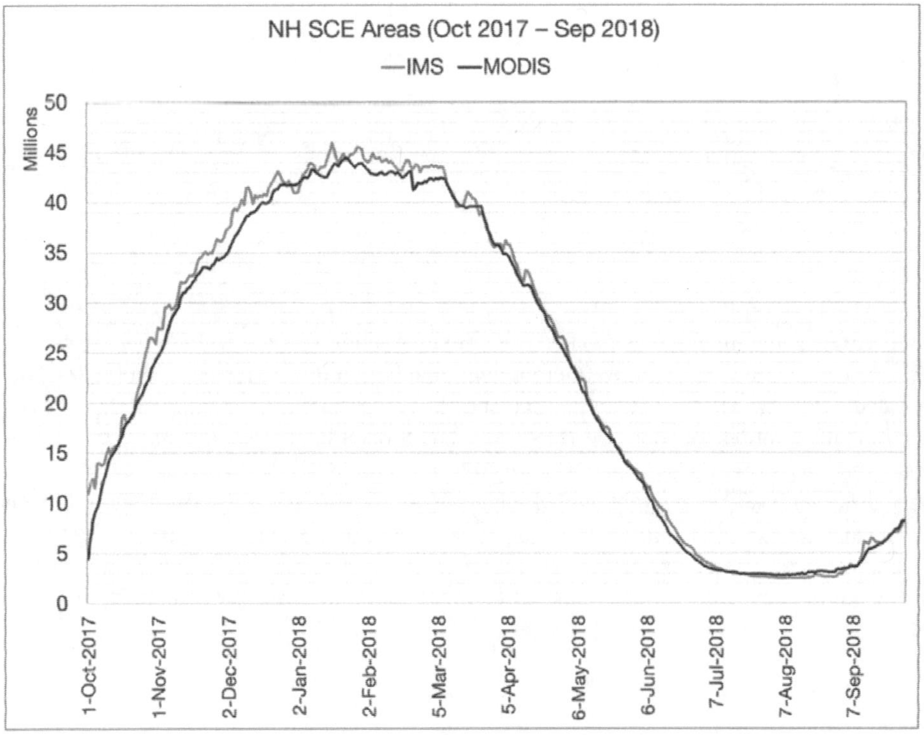

Figure 7. Same as figure 6 except for October 2017 to September 2018.

With, by far, the most apparent impediment to comparisons removed, we next delve into more detailed comparisons. This includes comparing monthly SCE areas for each product from the 2010-/2011 to 2019/2020 seasons. Such an evaluation for Octobers finds the IMS product depicting more cover in most years (Fig. 8). Differences range from the areas being within a few percent in four years and up to 12% different in six years. November differences (not shown) lean the same way but do not exceed 7%.

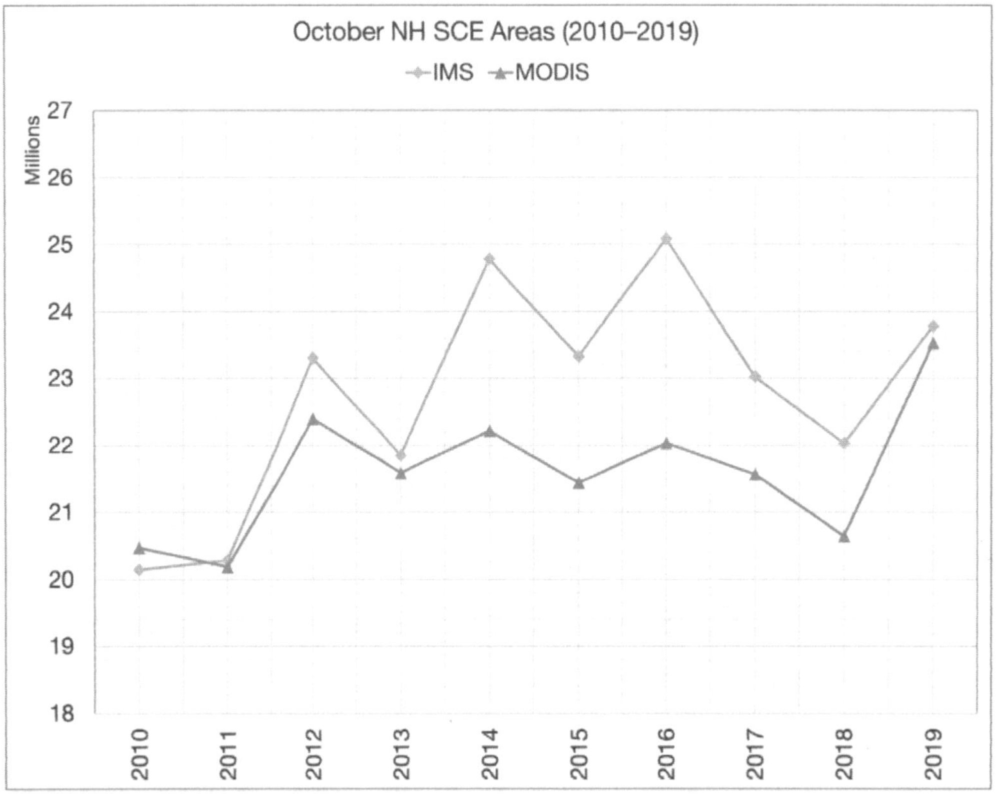

Figure 8. MODIS and IMS October Northern Hemisphere SCE from 2010 to 2019.

Spatial comparisons of the two products are next investigated. Figure 9 shows maps for four consecutive October days in 2016, the October with the largest difference between the MODIS and IMS products over the ten-year study period. The IMS maps show more snow on day one of each comparison but on subsequent days the IMS and MODIS maps begin to compare more closely. This was also noted in November, as seen in Figure 10 from 2010. Further comparisons are needed using ancillary data, but it is apparent that the IMS analysts are able to identify snow cover more quickly than in the automated MODIS approach, with the latter quickly "catching up" to the IMS coverage, though complete depiction of cover due to persistent cloudiness remains a mapping challenge. While having an impact on daily comparisons, differences are minimized when considering monthly values.

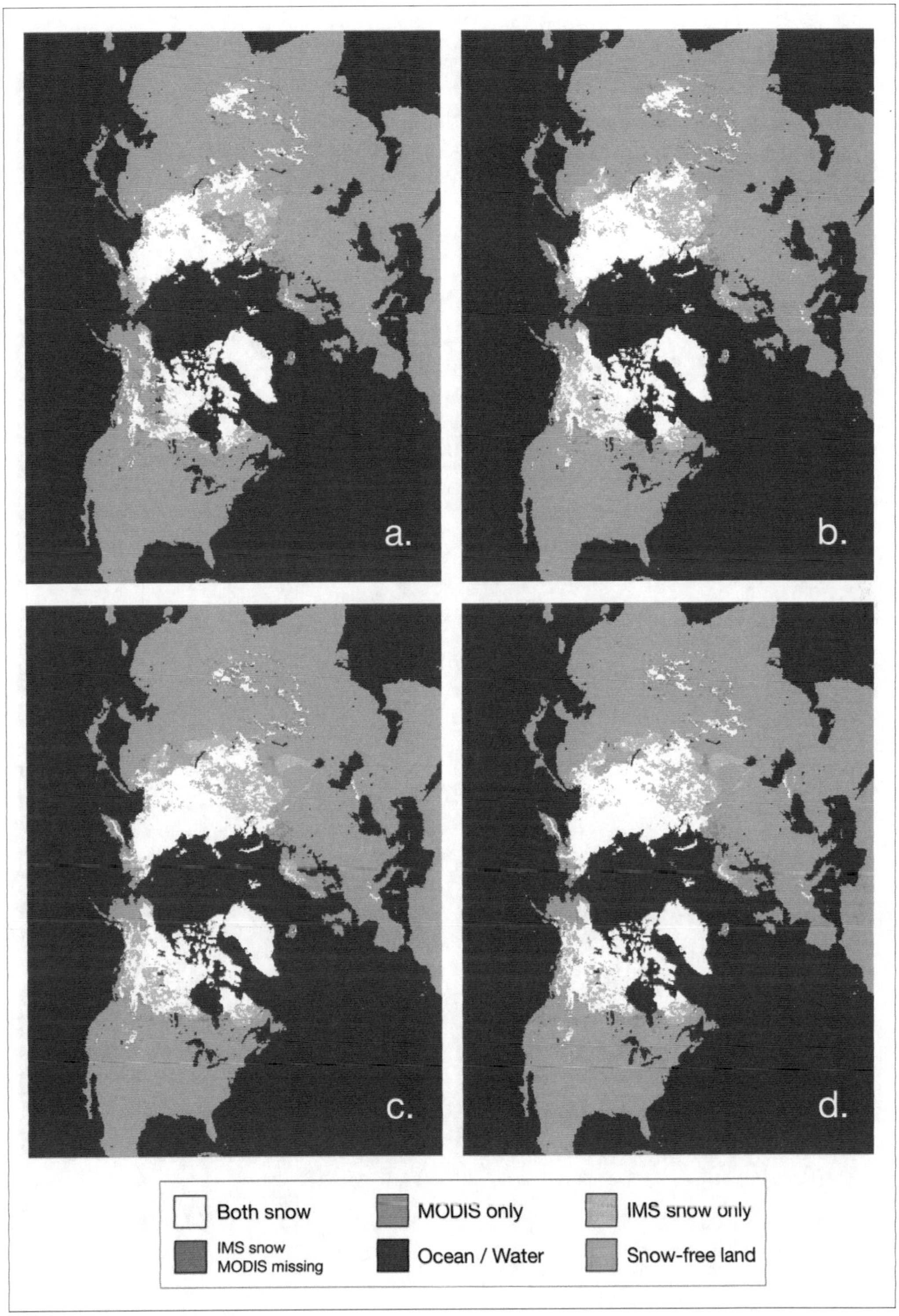

Figure 9. Example of maps comparing the two snow products on consecutive days: (a) 16 October 2016, (b) 17 October 2016, (c) 18 October 2016, and (d) 19 October 2016.

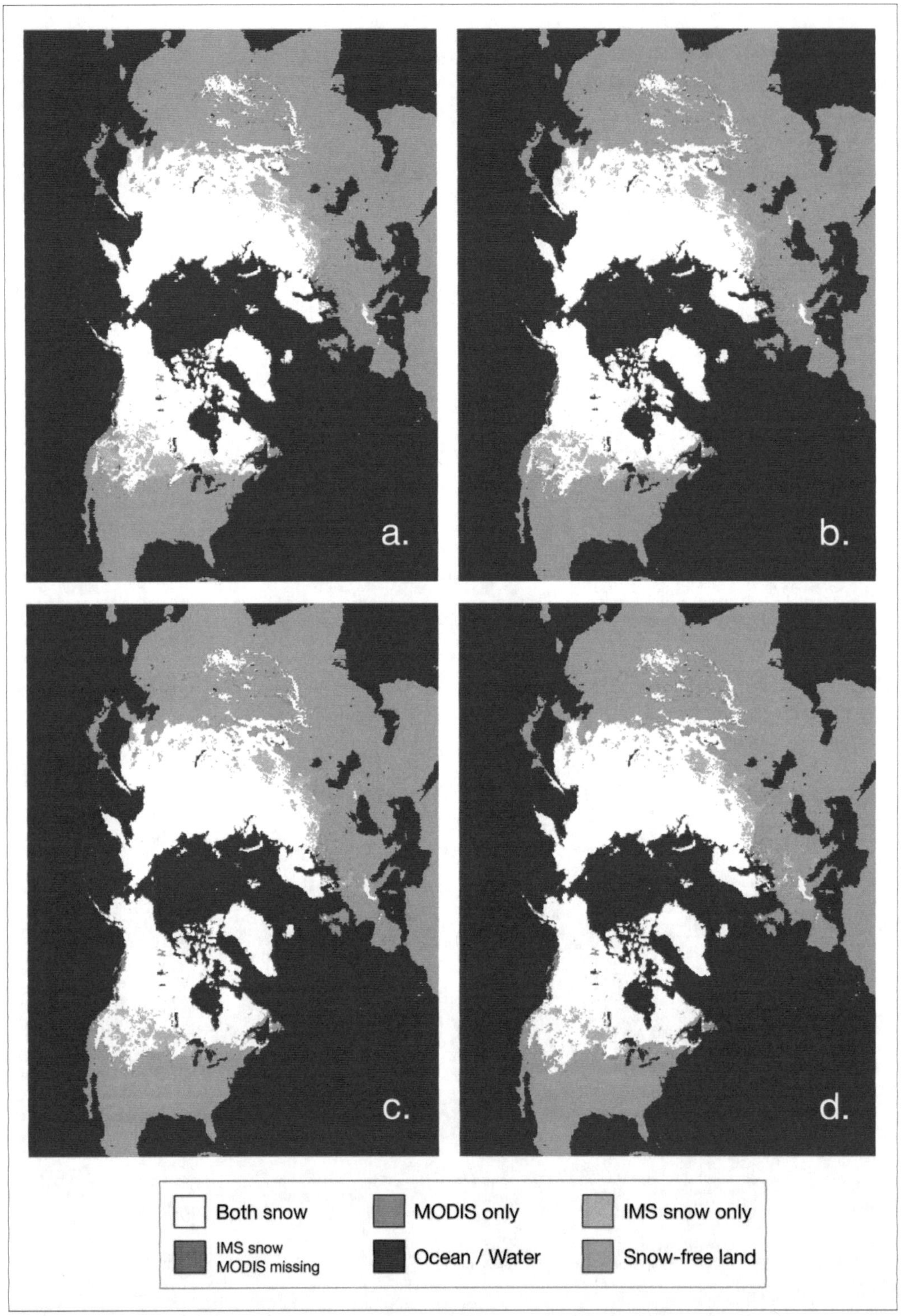

Figure 10. Example of maps comparing the two snow products on consecutive days: (a) 22 November 2010, (b) 23 November 2010, (c) 24 November 2010, (d) 25 November 2010.

DATA

Continuing with a fall focus, a comparison is made between the two products by mapping the monthly mean number of days over the 18-year study period where each showed the ground to be snow covered. An example is shown for November from 2002 to 2019 (Fig. 11). A large level of agreement is evident by the reasonable depiction of SCE for this early-season month. This approach also provides information on areas where there were disagreements between the products and the frequency of such differences. This is shown with two maps in Figure 12, one depicting where and how often snow is mapped in the MODIS product and not IMS and the other where IMS reports snow cover and MODIS does not. In both situations the average disagreement is generally no more than several days. Only in portions of Scandinavia do the MODIS maps show upwards of a week or more greater frequency of snow cover than IMS. Elsewhere, this tendency is mostly seen in mountainous areas of south-central Asia and western North America.

A different picture emerges where IMS reports cover and MODIS does not. The spatial coverage of this tendency is greater, covering the southern border of the snow boundary on both continents but with a lower frequency of days than found more locally in the greater MODIS snow cover locations. Further investigation into these patterns is needed using ancillary data. In the Scandinavian region it could be due to snow cover coming and going in November, with cloud persistence and product resolution differences impacting both products. The zonal difference might be more a result of the previously described multi-day MODIS "catch up" situation.

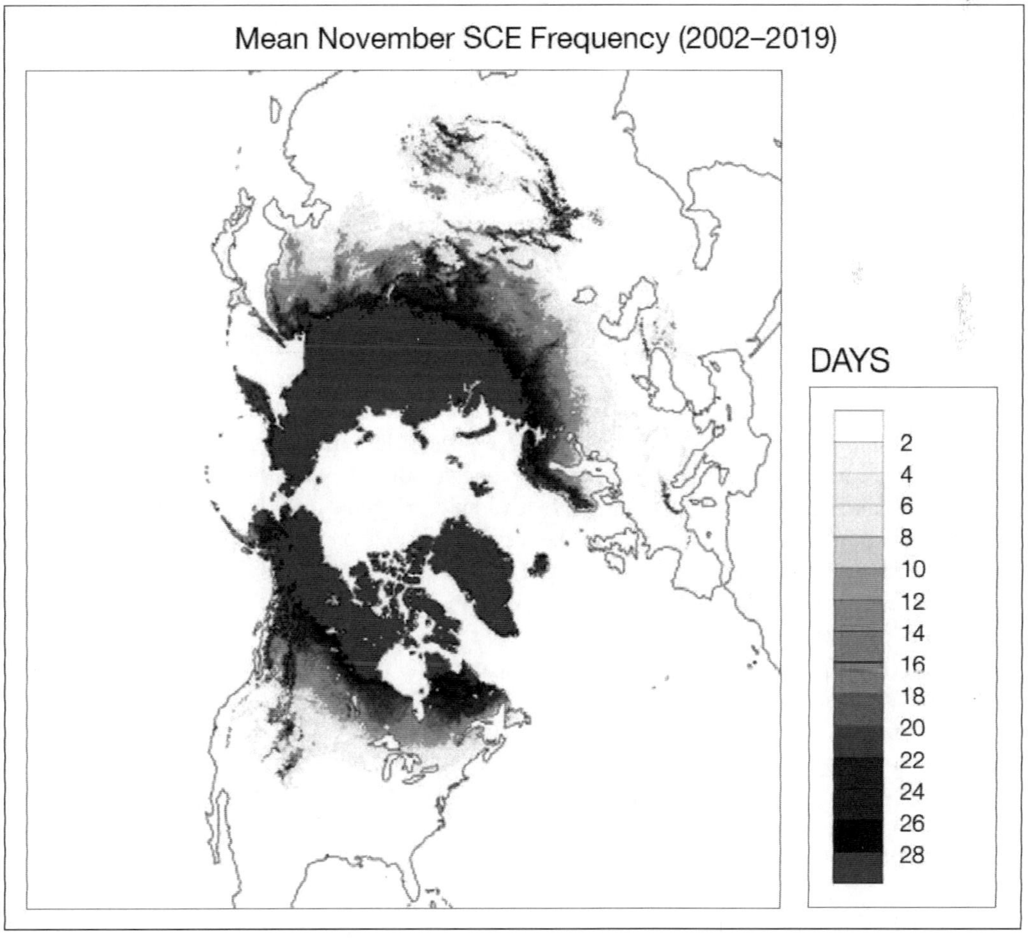

Figure 11. Mean November SCE frequency (2002–2019) where both MODIS and IMS show snow cover.

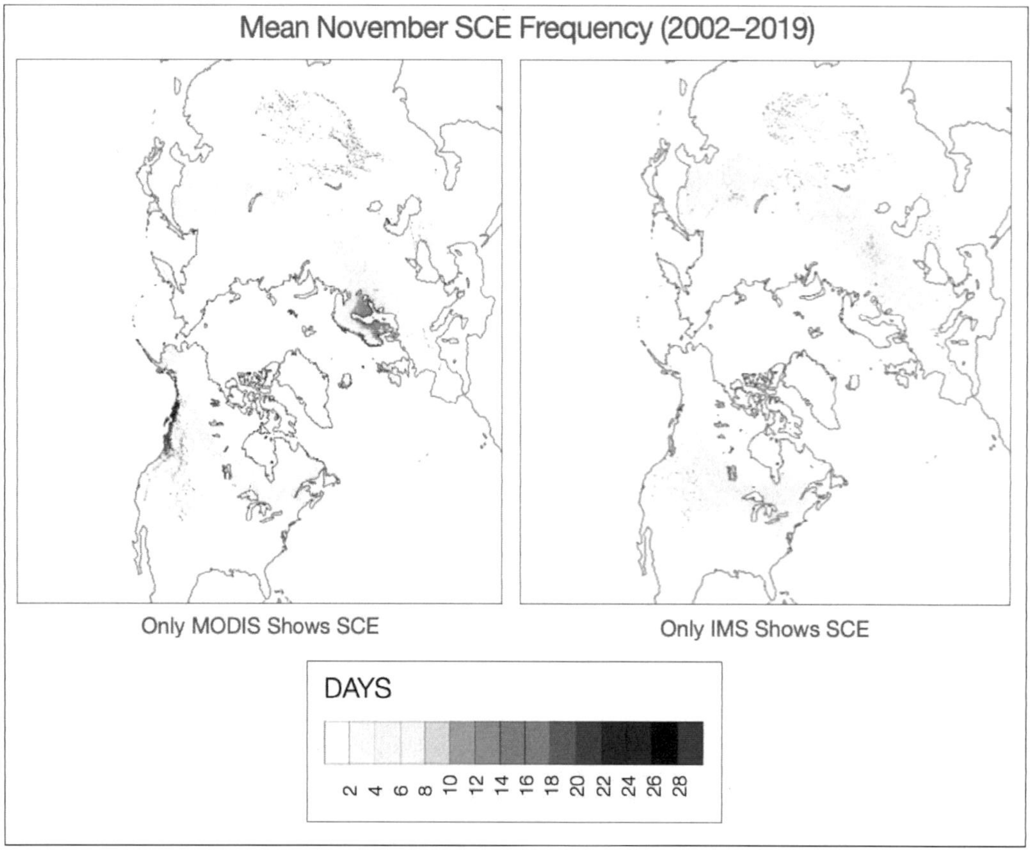

Figure 12. Mean November SCE frequency (2002-2019) where only MODIS shows snow cover (left) and only IMS shows snow cover (right).

Attention is next given to the spring melt season. As discussed earlier, the two products agree better during the melt season than in fall and winter. An examination of May SCE over the continents shows that all but one year of the ten show differences of less than approximately 0.5 million sq. km or several percent (Fig. 13). Four springs are essentially identical, two have MODIS showing more snow, and four with IMS mapping more. Examples of spatial differences between the products are shown for selected May days in 2017, the year with the largest difference between the two and 2019 where the monthly areas are equivalent (Fig. 14). In 2017, IMS SCE exceeds MODIS SCE at several locations along the southern periphery of the snow cover, in central Alaska, and some mountainous areas of south-central Asia. There are some quite localized areas where MODIS shows snow where IMS does not. The 2019 map shows very little spatial differences in cover between the two products. One small area along the snowline at lower elevations on each continent shows more snow mapped by IMS than MODIS, while locations where only MODIS mapped snow are more common in mountainous areas.

CONCLUSION

Temporal and spatial comparisons of 18 seasons of snow cover extent across Northern Hemisphere lands mapped by NASA MODIS and NOAA IMS snow cover extent products show an overall strong agreement on a macroscale across Northern Hemisphere continents. These results are achieved once masking of areas of low solar illumination in the MODIS product are removed, assuming snow cover in those high-latitude regions. A better agreement is noted in spring than fall perhaps because at the end of the winter season both have a common base level to start the melt season, while in fall there may be multiple episodes of snowfall and snowmelt accompanied by extensive cloudiness. The latter, in particular, appears to result in the manually generated IMS product recognizing snow cover a day or so before the automated MODIS product being more sensitive to cloud cover, catching up. Resolution issues with the IMS product in local areas, particularly with ephemeral snow cover and extensive fall cloudiness may impact accurate snow mapping.

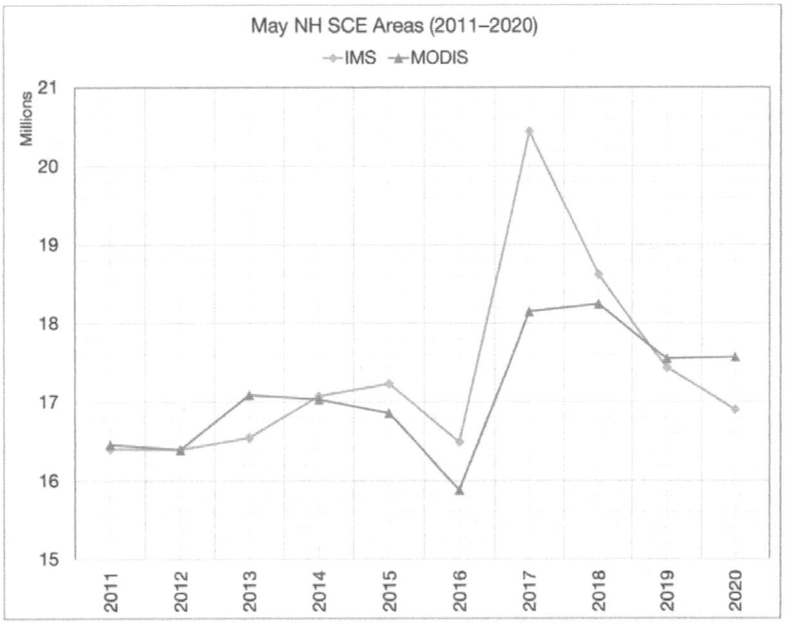

Figure 13. MODIS and IMS May Northern Hemisphere SCE from 2011 to 2020.

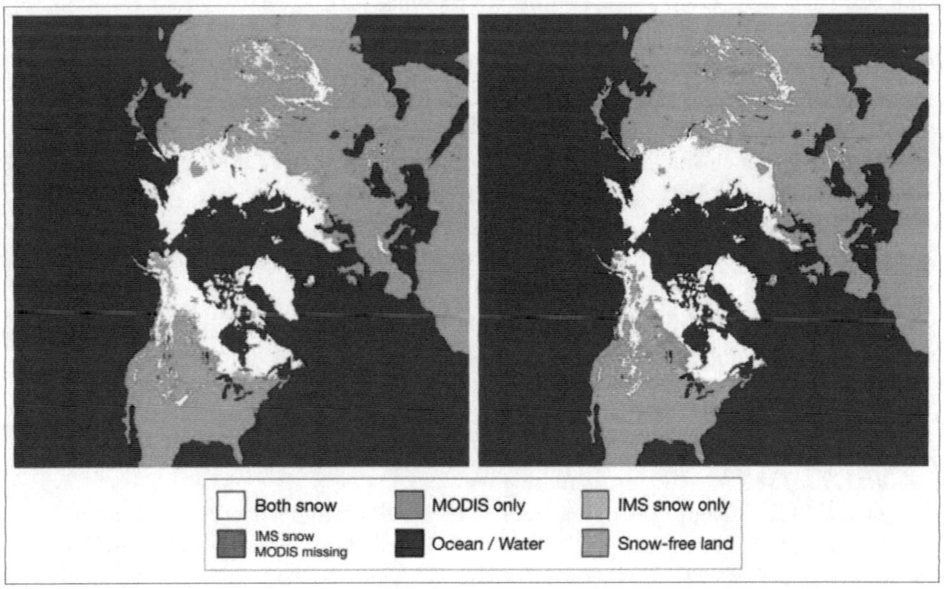

Figure 14. May 2, 2017 map comparing MODIS and IMS SCE (left) and May 2, 2019 map comparing the two products (right).

Ongoing comparisons include longitudinal evaluations over a wide range of landscapes and case studies for individual regions and specific dates. These will assist in confirming or refuting the assumption that snow cover indeed can most always be found in areas deemed in darkness by the MODIS methodology, provide a better understanding of intra- and inter-annual differences in SCE results between each product, and look specifically at potential local/regional differences in coastal and mountainous locations. This will involve using Landsat imagery and ground truth where station data are available. More robust statistical analyses will accompany these efforts.

Both CDR/ESDR products are currently archived at the National Snow and Ice Data Center (MODIS: *https://doi.org/10.5067/MODIS/MOD10A1F.061*, IMS: *https://doi.org/10.7265/N52R3PMC*). This will continue, with our team augmenting the documentation of each product with the knowledge gained through this study. Results of this project will provide user communities with the best understanding to date of the areal distribution of snow cover on regional to hemispheric scales for each product. Eventually, unifying these products will improve and extend the MEaSUREs snow product that currently extends only through 2012 (Robinson et al., 2014).

Acknowledgements. This study is supported by NASA funds to Rutgers University. The authors would also like to acknowledge support from NASA grant number 80NSSC21K1927.

REFERENCES

Allchin MI, Déry SJ. 2020. The climatological context of trends in the onset of Northern Hemisphere seasonal snow cover, 1972–2017. *Journal of Geophysical Research: Atmospheres*, **125:** e2019JD032367. doi: 10.1029/2019JD032367.

Estilow TW, Young AH, Robinson DA. 2015. A long-term Northern Hemisphere snow cover extent data record for climate studies and monitoring. *Earth System Science Data*, **7**(1):137-142. doi: 10.5194/essd-7-137-2015.

Hall DK, Riggs GA, Foster JL, Kumar SV. 2010. Development and evaluation of a cloud-gap-filled MODIS daily snow-cover product. *Remote Sensing of Environment*, **114**(3): 496-503. doi: 10.1016/j.rse.2009.10.007.

Hall DK, Riggs GA, DiGirolamo NE, Román MO. 2019. Evaluation of MODIS and VIIRS cloud-gap-filled snow-cover products for production of an earth science data record. *Hydrology and Earth System Sciences*, **23:** 5227–5241. doi: 10.5194/hess-23-5227-2019.

Hall DK, Riggs GA. 2020. MODIS/Terra CGF Snow Cover Daily L3 Global 500m SIN Grid, Version 61. Boulder, Colorado USA. NASA National Snow and Ice Data Center Distributed Active Archive Center. doi: 10.5067/MODIS/MOD10A1F.061.

Helfrich SR, McNamara D, Ramsay BH, Baldwin T, Kasheta T. 2007. Enhancements to and forthcoming developments in the Interactive Multisensor Snow and Ice Mapping System (IMS). *Hydrological Processes*, **21**: 1576–1586. doi: 10.1002/hyp.6720.

Riggs GA, Hall DK, Román MO. 2017. Overview of NASA's MODIS and Visible Infrared Imaging Radiometer Suite (VIIRS) snow-cover earth system data records. *Earth System Science Data*, **9**(2): 765-777. doi: 10.5194/essd-9-765-2017.

Riggs GA, Hall DK, Román MO. 2018. MODIS snow products user guide for Collection 6.1 (C6.1). Available at: *https://modis-snow-ice.gsfc.nasa.gov/?c=userguides*, last accessed 3/17/2019.

Riggs GA, Hall DK. 2020. Continuity of MODIS and VIIRS snow cover extent data products for development of an Earth Science Data Record. *Remote Sensing*, **12**, 3781. doi: 10.3390/rs12223781.

Robinson DA, Hall DK, Mote TL. 2014. MEaSUREs Northern Hemisphere Terrestrial Snow Cover Extent Daily 25km EASE-Grid 2.0, Version 1. Boulder, Colorado USA. NASA National Snow and Ice Data Center Distributed Active Archive Center. doi: 10.5067/MEASURES/CRYOSPHERE/nsidc-0530.001.

Robinson DA, Dewey KF, Heim Jr R. 1993. Global snow cover monitoring: an update. *Bulletin of the American Meteorological Society*, **74**: 1689–1696. doi: 10.1175/1520-0477(1993)074<1689:GSCMAU>2.0.CO;2.

Robinson DA, Dewey KF. 1990. Recent secular variations in the extent of Northern Hemisphere snow cover. *Geophysical Research Letters*, **17**: 1557–1560. doi: 10.1029/GL017i010p01557.

U.S. National Ice Center. 2008. IMS Daily Northern Hemisphere Snow and Ice Analysis at 1 km, 4 km, and 24 km Resolutions, Version 1. National Snow and Ice Data Center (NSIDC), Boulder, CO. doi: 10.7265/N52R3PMC.

Woods J, Helfrich S. 2019. Interactive Multisensor Snow and Ice Mapping System (IMS) upgrades and improvements. *Proceedings of the 76th Eastern Snow Conference, Fairlee, VT.*

Evolution of Global Snow Cover – Analysis of 23 Years of DLR's Global SnowPack and Latest Processor Developments

SEBASTIAN RÖßLER[1] AND ANDREAS DIETZ[1]

ABSTRACT

As an essential climate variable, the area covered by snow should be recorded daily and with a sufficient spatial resolution. This is currently only achieved with medium-resolution optical remote sensing sensors. The Global SnowPack processor developed at DLR enables the daily derivation of a gap-free (without data gaps due to clouds or polar night) global snow cover in near real time by combining temporal and topographical interpolation methods. So far, the daily snow product from MODIS on Terra and Aqua has been processed, but now also the daily snow product from Suomi NPP VIIRS is included (Sentinel-3 OLCI will also be included in the near future). As a result, we now have a 23-year data set of global uninterrupted snow cover, which can be used both for trend analysis and for forecasting extreme hydrological events. For the determination of long-term trends, the accuracy and the duration of the time series are decisive, for the hydrological application rather the timeliness.

Our pixel-based trend analysis showed most significant developments in the full snow cover season throughout the full hydrological year (Figure 1; meteorological autumn to summer), where more than 30% of the area showed a significant trend of snow cover duration (two-thirds show a decrease). In addition, analyzes at catchment area level were also made. We will present these results and application examples.

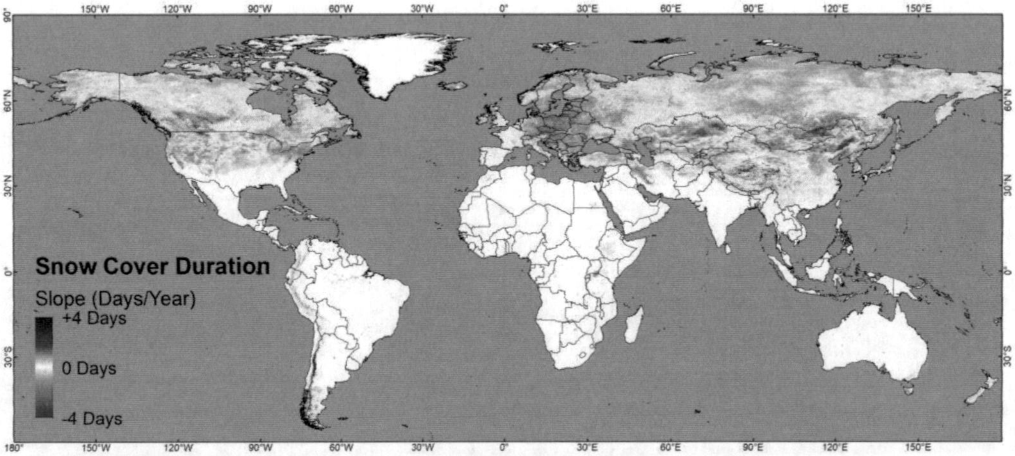

Figure 1. Global trend of full snow cover duration (whole hydrological year).

[1] German Aerospace Center, Oberpfaffenhofen, Germany
Corresponding author: sebastian_roessler@dlr.de

Characterization of the Vermont Snowpack

KATHERINE HALE[1], ANNA GRUNES[1], BEVERLEY WEMPLE[2], ARNE BOMBLIES[1], JAMES SHANLEY[3], AND ANDREW SCHROTH[2]

ABSTRACT

Across the northeastern United States, snowfall and the local to regional snowpack are critical components of the hydrologic cycle, yet their effect on the surrounding environmental system are relatively understudied. The most recent region-wide analyses of snow and winter-season climate in the northeast extend to the early 2000s, with low spatial coverage. Yet the northeastern snowpack plays a first principle role in delaying water availability to the terrestrial system by storing water until periods of intermittent and continuous snowmelt. Further, the amount of water in the snowpack and the timing of snowmelt are expected to strongly influence downstream runoff (timing and volume), spring nutrient fluxes, flooding, and drought. Both snow water equivalent (SWE) and snowpack water storage representation have largely been overlooked in the northeast, in part due to scarce observations. In response, a cold-weather environmental observation network has been established to span the summit-to-shore continuum in Vermont for advanced environmental analysis and planning and operational decision-making capability. The 22 monitoring stations span from near Lake Champlain in Burlington, VT, to a high-elevation station near the summit of Mount Mansfield and extends to the Sleepers River research Watershed in Danville, VT to incorporate and build upon existing longer-term records of snow observations. Using distributed ground-based sensors combined with unmanned aircraft system (UAS)-based sensing and hydrologic modeling, this effort provides an improved understanding and characterization of snowpack spatial and temporal variability across aspects, elevations, and canopy coverage regimes.

[1] Department of Civil and Environmental Engineering, University of Vermont, Burlington, VT, USA
[2] Department of Geography and Geosciences, University of Vermont, Burlington, VT, USA
[3] New England Water Science Center, United States Geological Survey, Montpelier, VT, USA
Corresponding author: khale1@uvm.edu

Canadian Snow Radar Satellite Mission Science Readiness Advancements

BENOIT MONTPETIT[1], JOSHUA KING[1], CHRIS DERKSEN[1], RICHARD KELLY[2], PAUL SIQUEIRA[3], PETER TOOSE[1], AARON THOMPSON[2], ALEXANDRE ROY[4], ALEXANDRE GELINAS[4], ALEXANDRE LANGLOIS[5], JULIEN MELOCHE[5], VINCENT VIONNET[6], AND ANNA WENDLEDER[7]

ABSTRACT

Environment and Climate Change Canada (ECCC) and the Canadian Space Agency (CSA) continue to advance a new satellite Ku-band radar mission focused on providing moderate resolution (500 m) information on seasonal snow mass. Like many regions of the northern hemisphere, estimates of the amount of water stored as seasonal snow are highly uncertain across Canada. To address this gap, a technical concept capable of providing dual-polarization (VV/VH), moderate resolution (500 m), wide swath (~250 km), and high duty cycle (~25% SAR-on time) Ku-band radar measurements at two frequencies (13.5; 17.25 GHz) is under development. In this presentation, results from the Trail Valley Creek experiment (TVCEx) conducted in winter of 2018-19 will be presented. Data collected during the CryoSAR 2022-23 campaign in Powassan, Ontario, Canada will also be shown in the context of how the proposed snow radar mission can improve SWE retrievals in agricultural lands.

Using the UMASS airborne Ku-band radar instrument and satellite observations from RADARSAT-2 and TerraSAR-X, we show that it is possible to retrieve background soil properties allowing to separate the background from the snowpack contribution of the Ku-band signal and isolate the snow volume scattering to facilitate radar-based SWE retrievals. We also show that the ground-based snow sampling strategy deployed during the TVCEx, providing statistical distributions of snow microstructure and density, is crucial to properly estimate the radar signal from forward modelling.

Ground-based snow properties, soil and weather station information, drone LiDAR/optical data, and radar observations collected for the CryoSAR campaign will also be presented.

---

[1] Environment and Climate Change Canada, Climate Research Division, Ottawa/Toronto, ON, Canada.

[2] University of Waterloo, Department of Geography and Environmental Management, Waterloo, ON, Canada.

[3] University of Massachusetts Amherst, Department of Electrical and Computer Engineering, Amherst, MA, USA.

[4] Université du Québec à Trois-Rivières, Département des Sciences de l'Environnement, Trois-Rivières, QC, Canada.

[5] Université de Sherbrooke, Département de Géomatique Appliquée à l'Environnement, Sherbrooke, QC, Canada.

[6] Environment and Climate Change Canada, Meteorological Research Division, Dorval, QC, Canada.

[7] German Aerospace Center (DLR), German Remote Sensing Data Center, Oberpfaffenhofen, Germany.

Corresponding author: benoit.montpetit@ec.gc.ca

SNO-FOO 2023 – Peter Toose

The 2023 SNO-FOO award is presented to Peter Toose, scientist at Environment and Climate Change Canada.

It was pure coincidence that during his presentation, Miles Ecclestone mentioned that, compared to the early days of using drones in the field, they were now very easy and efficient to collect imagery and snow data. After a full winter campaign in Powassan, Ontario, Canada, calibrating and planning flight lines, Peter Toose had shown his great skills as a drone pilot collecting aerial photography for structure from motion method and LiDAR data for snow depth retrievals. Did I mention that this was an agricultural field with not too many obstacles? After all the science flights were completed, Peter decided to do some quick flights to collect some visual footage to document the field campaign. That's where he neglected to account for the height of the only small patch of trees in the field… The drone crashed. I would thus say that the drones are as efficient as their operator, Miles!

Many people have crashed drones in the field in way easier conditions than what Peter has flown. This alone would not be enough to get awarded the SNO-FOO award. On top of crashing the drone, Peter documented his last fight with much detail and posted it on the campaign's Slack channel for every participant in the campaign to witness. It's one thing to crash a drone but it's another to publicize it! Let it be a lesson to all young (and not so young) snow scientists out there: Be careful what you publish on the web, it might earn you the infamous SNO-FOO award!

Presented by Benoit Montpetit

Past Recipients

The Eastern Snow Conference annually bestows upon a distinguished snow scientist who, in striving for excellence in snow research, contributes to an event of notable humor the highly coveted Sno-Foo Award.

Year	Winner	Affiliation
2023	Peter Toose	Environment and Climate Change Canada
2019	Alexandre Roy	Université du Québec à Trois-Rivières
2018	Alexandre Langlois	Université de Sherbrooke
2017	Michael Durand	The Ohio State University
2016	Roger De Roo	University of Michigan
2015	Kevin Côté	Université de Sherbrooke
2014	Dorothy Hall	NASA GSFC
2013	Benoit Montpetit	Université de Sherbrooke
2012	Don Pierson	NYC DEP
2011	Ken Rancourt	Mt. Washington Observatory
2010	Kyung-Kuk (Kevin) Kang	University of Waterloo
2009	Rob Hellström	Bridgewater State University
2008	Steven Fassnacht	Colorado State University
2007	The Group of 9[1]	
2006	Andrew Klein	Texas A&M University
2005	Claude Duguay	University of Alaska-Fairbanks
2004	Chris Derksen	Meteorological Service of Canada
2003	Miles Ecclestone	Trent University
2002	Danny Marks	U.S.D.A.
2001	Brenda Toth	University of Saskatchewan
2000	Mauri Pelto	Nichols College
1999	Ross Brown	Meteorological Service of Canada
1998	Mary Albert	CRREL
1997	Jean Stein	Université de Laval
1996	Colin Taylor	Trent University
1995	Mike Demuth	NHRI
1994	Bert Davis	CRREL
1993	John Pomeroy	NHRI
1992	Tom Niziol	National Weather Service
1991	Terry Prowse	NHRI
1990	Kersi Davar	University of New Brunswick
1989	Gerry Jones	INRS-EAU
1988	Robert Sykes	SUNY
1987	John Metcalfe	Meteorological Service of Canada
1986	Peter Adams	Trent University
1985	Don Wiesnet	National Weather Service
1984	Barry Goodison	Meteorological Service of Canada

[1] Jimmy MacDonald (University of Saskatchewan), Bill Floyd (University of British Columbia), Chris DeBeer (University of Saskatchewan), Wendell Koning (Government of Alberta), Jaime Hood (University of Calgary), Dankia Muir (University of Calgary), John Jackson (University of Calgary), Sarah Forte (University of Calgary), Masaki Hayashi (University of Calgary)